U.S. Marine Corps Combat Water Survival

Plus

Marine Rifle Marksmanship

Department of Defense

Pentagon Publishing

MCRP 3-02C

Marine Combat Water Survival

U.S. Marine Corps

PCN 144 000069 00

To Our Readers

Changes: Readers of this publication are encouraged to submit suggestions and changes that will improve it. Recommendations may be sent directly to Commanding General, Marine Corps Combat Development Command, Doctrine Division (C 42), 3300 Russell Road, Suite 318A, Quantico, VA 22134-5021 or by fax to 703-784-2917 (DSN 278-2917) or by E-mail to **morgann@mccdc.usmc.mil**. Recommendations should include the following information:

- Location of change
  - Publication number and title
  - Current page number
  - Paragraph number (if applicable)
  - Line number
  - Figure or table number (if applicable)
- Nature of change
  - Add, delete
  - Proposed new text, preferably double-spaced and typewritten
- Justification and/or source of change

Additional copies: A printed copy of this publication may be obtained from Marine Corps Logistics Base, Albany, GA 31704-5001, by following the instructions in MCBul 5600, *Marine Corps Doctrinal Publications Status*. An electronic copy may be obtained from the Doctrine Division, MCCDC, world wide web home page which is found at the following universal reference locator: **http://www.doctrine.usmc.mil**.

Unless otherwise stated, whenever the masculine gender is used, both men and women are included.

DEPARTMENT OF THE NAVY
Headquarters United States Marine Corps
Washington, D.C. 20380-1775

6 January 2003

FOREWORD

Marine Corps Reference Publication (MCRP) 3-02C, *Marine Combat Water Survival*, provides Marine Corps combat water survival techniques, procedures, and training standards. This publication also teaches Marines to cross water obstacles and perform water rescues correctly and safely.

This publication is the foundation for teaching Marines correct water survival techniques and procedures that are used throughout the Marine combat water survival program (MCWSP). Once an individual or a unit has completed the MCWSP, this publication can be used as a refresher course before water operations.

MCRP 3-02C supersedes Fleet Marine Force Reference Publication (FMFRP) 0-13, *Marine Combat Water Survival*, dated 16 September 1991.

Reviewed and approved this date.

BY DIRECTION OF THE COMMANDANT OF THE MARINE CORPS

EDWARD HANLON, JR.
Lieutenant General, U.S. Marine Corps
Commanding General
Marine Corps Combat Development Command

Publication Control Number: 144 000069 00

MCRP 3-02C Marine Combat Water Survival

Table of Contents

Chapter 1. Survival at Sea

Chapter 2. Water Rescues

Chapter 3. Treatment of Casualties and Avoidance of Dangerous Marine Life

Chapter 4. Negotiating Water Obstacles

Chapter 5. Fording Waterways

Appendix. Knot Tying

Throughout history, water has posed special challenges to Marines and Sailors during times of both peace and war. Therefore, the inherent dangers associated with waterborne operations demand that Marines and Sailors receive proper water survival training. Combat units that are confident in their ability to work in and around water can use the water to their advantage in combat, and history is filled with examples of how the proper preparation or training for survival in water reduced or averted disaster.

USS *Indianapolis*

On Sunday, 29 July 29 1945, the heavy cruiser USS *Indianapolis* (the *Indy*) was en route to the Philippine Sea. Shortly before midnight (about 39 hours out of port), the *Indy*, running blacked out and unescorted, was rocked by two explosions on her starboard side. With communications smashed, the ship could not signal its distress and sank within 15 minutes.

Three life rafts and a floater net supported a few survivors, but the rest of the survivors drifted about, held up by rubber life belts or Mae Wests. About 60 seamen died the first night. Survivors assumed the ship would be reported overdue in Leyte, and that they would be rescued within 2 days. By Monday evening, however, panic began to set in as some life jackets lost their buoyancy from the long immersion. Some men even fought over life jackets, which resulted in at least 25 deaths. No one dared sleep for fear of losing his jacket.

Throughout the next several days, in-transit aircraft flew nearby without spotting the desperate seamen. As best they could, the men kept together, some tying long ropes to each other, floating like corks on a net.

Not until late Thursday morning, 3 1/2 days after the ship sank, were the men discovered. Luckily, a plane on a routine flight over the area spotted the survivors. When ships picked them up that night, the survivors learned they had never been reported overdue. Every one of the 1,196-man crew was a casualty; 880 crewmembers were listed as dead or missing. Although many lives were lost, the innovative and expedient use of flotation devices and float techniques employed by the survivors helped save hundreds of lives.

USS *America*

On Thanksgiving eve, 23 November 1995, the USS *America* made its way through the Arabian Sea. Twenty-year-old, Marine Lance Corporal (LCpl) Zachary Mayo was unable to sleep and, wanting some fresh air, made his way onto an open-air platform near the aircraft hangar bay, which was three levels below the sleeping quarters. While he was on the platform, the ship veered suddenly, throwing LCpl Mayo through the platform's protective bars and into the sea, 30 feet below.

Frantic, LCpl Mayo called out in vain to the watchmen on the flight deck, which was 64 feet above him. It soon became clear to him that the USS *America* would keep its course into the Gulf of Oman until his absence was discovered at morning muster. The LCpl took a moment to consider his situation. Since land was at least 100 miles away, swimming was suicide; he would have to stay afloat until a search party found him.

Using the techniques he had learned during combat water survival training, LCpl Mayo made a flotation device out of his coveralls and tried to relax. Meanwhile, business continued as usual aboard the USS *America*. Since LCpl Mayo was on special assignment with the hazardous materials division, his absence went unnoticed until a petty officer asked several Sailors if they had seen their shipmate recently. By the time a roll call had been completed, LCpl Mayo had been adrift at sea for over 24 hours. Although three, fixed-wing Viking aircraft were deployed to search for LCpl Mayo, most people aboard ship feared the worst.

After 34 hours at sea, LCpl Mayo was discovered by fishermen on a Pakistani fishing boat. LCpl Mayo's survival is a testament not only to his incredible physical courage, but also to the soundness of the lifesaving training and techniques he received during combat water survival training.

Chapter 1
Survival at Sea

As a Marine, you face a variety of potential water emergencies whenever you cross expanses of water: ships, watercraft, and amphibious assault vehicles (AAVs) can sink; aircraft can crash into the sea; or you can accidentally fall into the water. However, there are some basic precautionary measures you can take to protect your safety and reduce your chances of becoming a water casualty. Determine the following information as soon as you board any type of vessel. Your knowing the following information may save your life or the life of your fellow shipmates.

- How many life preservers and lifeboats/rafts are on board?
- Where are the life preservers and lifeboats/rafts located?
- What type of unit survival equipment is on board?
- Are individual survival kits issued to each person on board?
- How much food, water, and medicine do the survival kits contain? When was the last time the contents were inspected for proper quantities and shelf life expiration?
- Is there sufficient survival equipment available for the number of personnel?
- How many other personnel are there on board, and where are they located?
- What are the egress procedures for the ship, boat, watercraft, AAV, or aircraft?

Abandoning Ship

When you embark on a Navy ship, you will receive abandoning ship instructions from Navy personnel. If given the order to abandon ship, report to your designated assembly area and put on a life preserver. DO NOT inflate the life preserver until you are clear of the ship. Torn life preservers will not inflate and inflated life preservers can block you, and those behind you, from exiting the ship. A flotation device that has been inflated may also burst if you jump from a significant height. See pages 1-15 through 1-35 for staying afloat with and without a life preserver.

DO NOT remove your clothing, boots, or shoes before abandoning ship. Your trousers and blouse may be the only flotation devices available if your life preserver is faulty or becomes damaged, and your clothes can provide some insulation from the cold water. However, remove your soft cover and place it in a cargo pocket for later use. The soft cover is both lightweight and a good protection against sunburn caused by the sun's rays reflecting off the water.

Jettisoning Equipment

Equipment should be kept properly packed and waterproofed in case you have to abandon ship. If entering the water from a height greater than 30 feet, wearing your equipment (e.g., pack, helmet, gas mask) could cause injury. Upon impact with the water, the helmet will "cup" air inside of it. The chin strap may also create a "hanging effect" as you submerge from the force of the fall. Therefore, you should remove your helmet and gas mask before abandoning ship.

If you are unable to maintain buoyancy due to the amount of equipment secured to your pack and body, then jettisoning some of your equipment may become necessary. Equipment that you should always retain includes canteens of freshwater, first aid kit, soft cover, and survival kit. The survival kit should include first aid items, water purification tablets or drops, fire starting equipment, signalling items (e.g., flashlight, strobe light, chemlights), food procurement items, and shelter items. Other items in a survival kit include sunburn lotion and lip balm, knife, goggles/sunglasses, plastic bag, matches and lighter, mirror. See Marine Corps Reference Publication (MCRP) 3-02F, *Survival*, for a detailed list of survival items and applications.

Abandoning Ship Technique

When abandoning ship, safety considerations must be observed. Use the following technique when abandoning ship without your combat gear:

Place your hands on their opposite shoulders, forming a criss-cross pattern.

Step to the edge of the ship's deck and check the water below for debris or survivors. If the water is clear, look straight ahead and prepare to jump. If the water is not clear, move to another location.

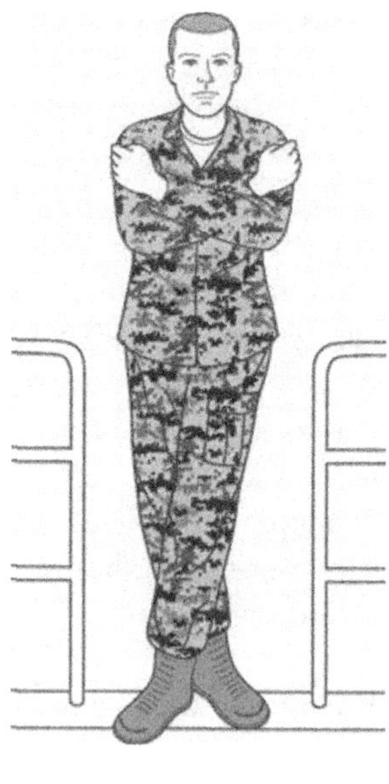

NOTE: DO NOT hold your nose as you abandon ship. If you do hold your nose, the force of impact into the water could jar your arm and hand and cause you to break your nose.

Step off the side of the ship with a smooth, 30-inch stride. DO NOT DIVE OFF THE SHIP. DO NOT LOOK DOWN AT THE WATER. LOOK STRAIGHT AHEAD. Looking down at the water can render you unconscious or cause injuries upon impact.

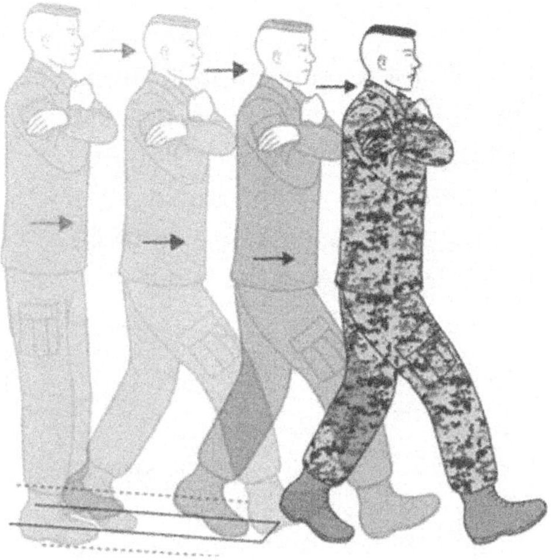

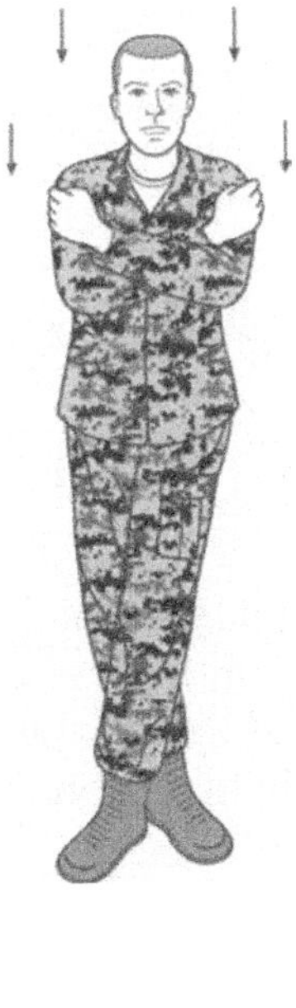

Bring your trailing leg forward during the fall. Cross your trailing leg behind your leading leg.

Keep your head parallel to the water's surface until hitting the water.

You should remain in the abandon ship position until your descent into the water has almost stopped. However, the weight imbalances in your body may cause you to be in a "J" shape under the water

.

Once your downward motion has ceased, your feet may be parallel with the ocean bottom or you may be nearly inverted with your feet over your head. To counteract potential disorientation, you should pause briefly and allow the natural buoyancy of your torso to bring your body to a nearly upright position.

Floating debris can cause hazards. Therefore, you should swim upward, extending one arm (hand is shaped as a fist) upward to feel for obstructions. If you encounter debris, try to push it away or surface in a different location.

Swim away from the ship. DO NOT LOOK BACK AT THE SHIP. Looking back could slow your movement away from the area. Remember, your objective is to leave the area as quickly as possible because—

- Equipment and debris may be falling from or spilling out of the ship.
- Additional casualties can occur if individuals abandoning the ship fall on top of swimmers already in the water.
- Swimmers close to the sinking ship may get pulled underneath the water by the suctioning effect of the ship as it goes under.

Modified Abandoning Ship Technique

When abandoning ship while wearing full combat gear (weapon, helmet, and a properly waterproofed pack), safety considerations must be observed. The modified abandoning ship technique is used while wearing full combat gear and exiting from a height less than 30 feet.

WARNING

When exiting from a height greater than 30 feet, remove your helmet and pack. Fasten the helmet to the pack or place it inside the pack before jettisoning. If jettisoning gear from the ship or aircraft, check the water below for survivors before throwing the gear forward of the intended jump area. Once in the water, retrieve your gear and swim out of the area.

Place your weapon over one shoulder, muzzle down, with the weapon parallel to your side.

Place your arm and hand along the weapon and hold it to your side.

Take your free hand and place it on top of your helmet to prevent neck/spinal injury from the force of the water pulling upward on the helmet as your body enters the water.

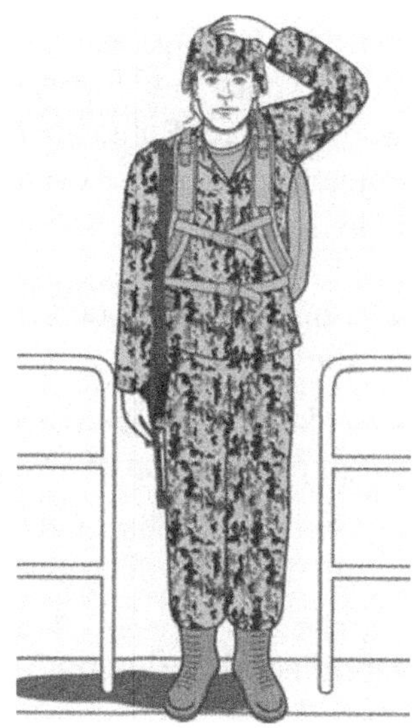

NOTE: An alternate method is to place your weapon over one shoulder, muzzle down, with the weapon parallel to your side. Reach across your body and grasp the sling of your weapon and hold it to your body. Take your free hand and place it on top of your helmet.

Step to the edge of the platform and check the water below for debris and survivors. If the water is clear, look straight ahead and prepare to jump. If the water is not clear, move to another location.

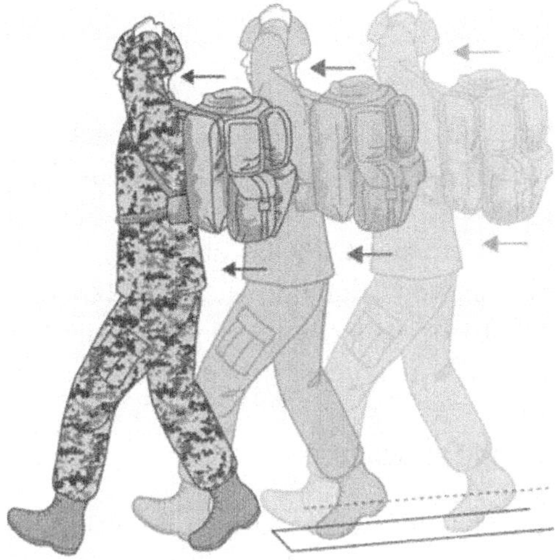

Step off the side of the platform with a smooth, 30-inch stride. DO NOT DIVE OFF THE SHIP, DO NOT LOOK DOWN AT THE WATER, LOOK STRAIGHT AHEAD. Looking down at the water can render you unconscious or cause injury upon impact.

Bring your trailing leg forward during the fall. Cross your trailing leg behind your leading leg.

Keep your head parallel to the water's surface until hitting the water.

You should remain in the modified abandoning ship position until your descent into the water has almost stopped. The buoyancy of a properly waterproofed pack will immediately pull you to the surface. Once you break the water's surface, unsling your weapon and loop the sling over your head. You may have to seesaw the sling so as to ensure the sling passes between your helmet and pack. Once over your head, loop the sling around your neck with the weapon aligned in the center of your body, muzzle down. Lean back on your pack and perform the combat travel stroke to exit the area.

> *Note:* You may also remove the pack (unsnap the quick-release strap on one side of the pack, and pass the strap to your other hand in order to maintain contact with the pack, which is functioning as your flotation device). Remain horizontal in the water with your pack under your chest. Perform a breast stroke kick.

Surface Burning Oil Swim

After you have abandoned ship, rise to the surface using the techniques shown on the page 1-6. However, you must remember that fuel from sinking ships or downed aircraft will float on the surface of the water. Therefore, you must move clear of the floating fuel by swimming away from the ship or aircraft as soon as possible. Either swim upwind (into the wind) of the ship/aircraft or swim against the current. Either method allows you to move away from the fuel and the wind/current will push the fuel past you. To properly execute a surface burning oil swim—

Extend your arms overhead as far as possible.

Wave your arms back and forth vigorously to splash a hole while moving upward.

Splash as long as possible to push burning fuel away from the surfacing area.

Use your arms and hands to sweep away fuel and debris.

Kick your legs in a constant breast stroke kick.

Extend your arms (palms outward) forward on the surface, arms shoulder-width apart.

Pull your hands in and back toward the chest.

Stop your hands in front of your face and rotate them so that your palms face forward (roughly halfway out of the water).

Sweep your arms forward to a full extension at the shoulder width. This splashes debris, oil, or burning liquids aside. To reduce the chance of fatigue, use two short splashes to the front to extend the path.

Repeat the preceding step as necessary while swimming clear of the area.

Surviving With a Pack

If packed properly, your pack will float, and it is your key piece of equipment for staying afloat and overcoming water obstacles. If the pack's contents are properly waterproofed, it can support you (with a combat load) in the water. Buoyed up by a waterproofed pack, you eventually emerge from the water with all your equipment (e.g., boots, helmet, flak jacket, weapon, survival items).

Your pack floats based on a scientific principle known as Archimedes' principle. This principle states that an object submerged in a liquid is buoyed up by a force equal to the weight of the liquid displaced (pushed aside) by the object. If the weight of the displaced liquid is greater than the weight of the object, the object floats. If the weight of the displaced liquid is less than the weight of the object, the object sinks. For example, a machine gun sinks in the water, but it still weighs less in the water than it does on land. Even though a machine gun sinks, it is still buoyed up by a force equal to the weight of the water it displaces. For this reason, you should not try to hold yourself or your equipment any higher out of the water than they would naturally float; doing so wastes both energy and body heat.

Preparing Equipment

Before packing your pack, you must prepare your gear/equipment. Tape or pad all sharp edges and equipment corners. Ideally, your gear/equipment is placed in plastic bags and the plastic bag is then placed inside the standard-issued, rubberized, waterproof bag. Waterproof bags are not completely water tight; henceforth, the added protection of first wrapping the gear/equipment in a plastic bag, then placing the plastic bag inside the waterproof bag. Large plastic bags (e.g., trash bags) work well for bulky equipment (e.g., sleeping bags, field jackets, shelter halves, gas masks). Small plastic bags work best for small items (e.g., shaving gear).

> *Note:* If a gas mask must be carried outside the pack, cover it with a waterproof bag.

You will need the following items, which are available through the supply system, to prepare your equipment for packing:

- Waterproof bags:
  Pistol bag, plastic size #1 (8" x 18").
  Rifle bag, plastic size #2 (10" x 56").
  Machine gun bag, plastic size #3 (15" x 56").
  Multipurpose bag, cover, plastic size #4 (20" x 84").

- Panama work vest (life preserver).
- Small plastic bag, self sealing (6" x 6" and 12" x 12").
- Riggers tape, olive drab (2 1/2").
- Willie peter bag, waterproof clothing.

Tying Waterproof/Plastic Bags

Try to remove the excess air from the waterproof/plastic bag before securing the bag's opening. If filled with air, the bag can burst if pressed from the outside. Perform the following steps to tie a waterproof/plastic bag:

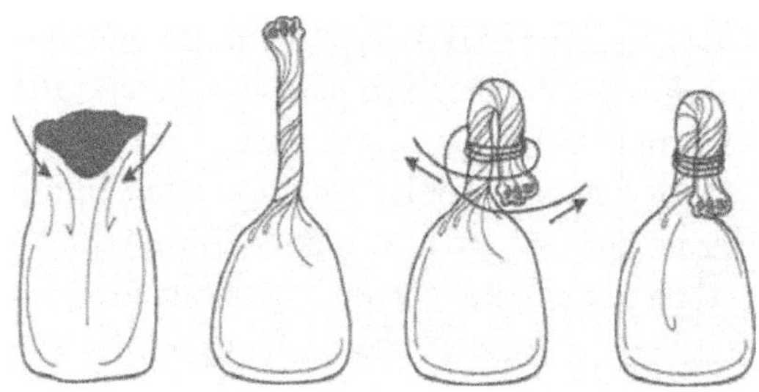

Packing the Pack

Place filled and tied bags inside the pack. Carefully handle sharp items (e.g., tent pegs, poles) to prevent puncturing the bag. Place items in the pack in order of expected use. Close the pack and its compartments. Attach sleeping mats or bags as high as possible on the outside of the pack.

Swimming With the Pack

A useful technique for propelling yourself forward while wearing a pack is the combat travel stroke. This technique allows you to float nearly horizontal and to propel yourself forward with bicycle-style kicks and breast stroke-style arm sweeps. To execute the combat travel stroke—

Body position
Keep the upper part of your body prone to the water, let your legs dangle horizontally, and keep your face up.

Arm action
Extend your hands out in front of your waist. Sweep your arms slightly downward and back 90 degrees to propel your body through the water. Move your hands back to the front of your waist. Repeat.

Leg action
Continuously move your legs in a bicycle-like movement, bringing your knees up high and step out.

Breathing
Keep your face out of the water during the stroke, and breathe freely.

Staying Afloat With a Life Preserver

The best form of flotation is to find any kind of floating object that will keep you and your equipment out of the water or minimize your exposure to the water. Life preservers are the best method, as they allow you to wear your clothes for heat retention and sunburn prevention. Marines use two basic classes of life preservers: inherently buoyant life preservers and inflatable life preservers.

Inherently Buoyant Life Preservers

Inherently buoyant life preservers are either vest-type (worn like a jacket) or yoke-type (worn around the neck). The preserver's outer envelope is either a cotton or water resistant material that encloses a removable fibrous glass or plastic foam filling.

In the Marine Corps, the most common type of inherently buoyant life preserver is the vest-type with collar, known as the kapock preserver. The kapock consists of collar straps, upper front chest straps, leg straps, and waist drawstrings that secure the preserver to you. The leg straps, which are fitted on both sides of the life preserver, ensure that the preserver remains around your chest while you are in the water. A chest strap is attached to the life preserver to facilitate lifting you out of the water. The strap can also be attached to other survivors or to lifeboats to reduce the fatigue that results from holding onto a floating/secured object by hand.

Inflatable Life Preservers

Marine Corps aircraft and AAVs have inflatable life preservers on board. The inflatable life preserver issued to Marines is known as the LPP (life preserver personal). The LPP is capable of both oral inflation and CO_2 cartridge inflation. The LPP consists of buoyancy chambers, CO_2 inflator, and an oral inflation tube. The buoyancy chamber is deck gray in color and is made from a neoprene-coated nylon fabric.

Inflatable life preservers must be stored in a cool, dry place. Heat, moisture, and light cause deterioration of the life preserver material. Do not stow CO_2 cylinders near steam lines or radiators. Heat can increase the pressure inside the cylinders causing them to explode. Avoid sharp edges in stowage. Sharp edges increase wear and tear on the life preservers and may also puncture inflatable buoyancy chambers.

> **CAUTION:** Do not inflate the life preserver until you are clear of the aircraft, ship, or vehicle. Torn life preservers will not inflate and inflated life preservers can block you, and those behind you, from exiting the aircraft, ship, or vehicle.

To don and adjust the LPP—

Remove the LPP from its storage container.

Fasten the belt fasteners in front with the pouch in the rear.

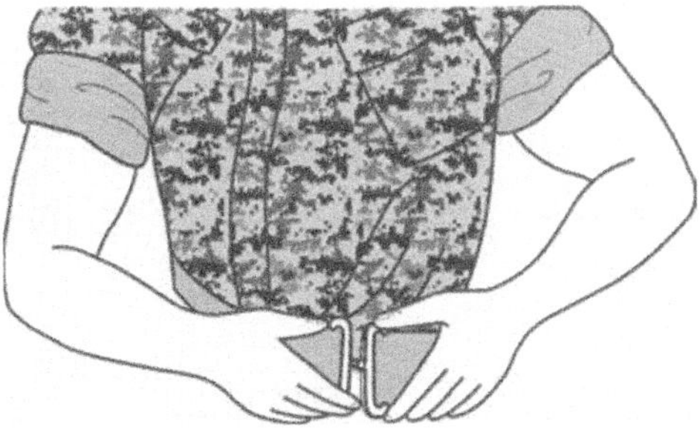

Adjust the belt to fit; secure any excess belt by mating the hook and pile tape.

Rotate the pouch to the front and re-adjust the belt if necessary.

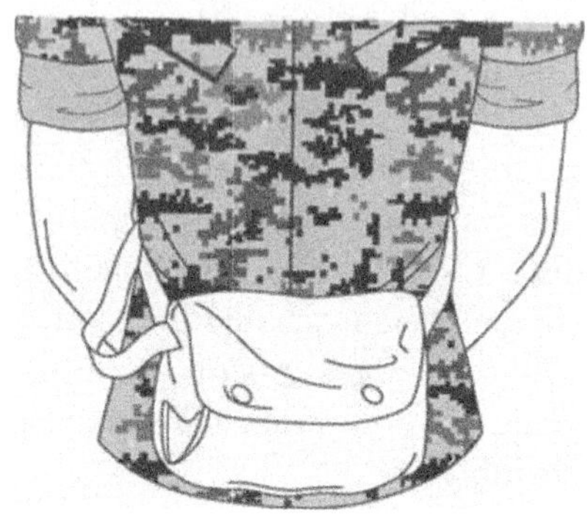

Open the snap fasteners on the pouch and unfold the life preserver.

Place the deflated life preserver over your head.

Place the storage container into the pouch after donning the life preserver.

Lift the lower end of the life preserver out of the pouch.

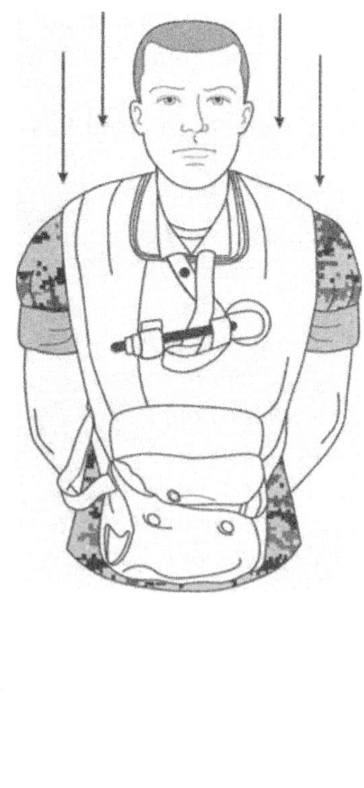

Inflate the life preserver by pulling on the lanyard attached to the CO_2 inflation valve or by blowing on the end of the oral inflation valve.

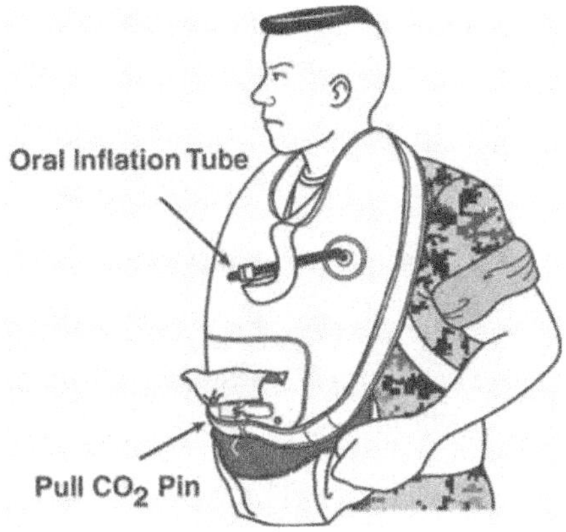

Staying Afloat Without a Life Preserver

You may be in open water without any floating objects or a life preserver to help you survive. If so, uniform blouses and trousers can be made into expedient flotation devices.

Floating With an Inflated Blouse

It is possible to float by a bubble of air trapped in the shoulders of your blouse. The air rises to the back and shoulders of the blouse and supports you at the water's surface. An inflated blouse is also a temporary flotation device used by weaker swimmers while trying to remove their trousers. There is a primary and an alternate way to create a bubble of trapped air in a blouse—

Primary Method
Turn the collar inside the blouse to help create a seal.

Unbutton top button and pull collar around mouth and nose.

Take a deep breath and bend forward slightly at the waist. Exhale one-half to three-quarters of a breath into the blouse.

Grasp and twist the collar with one hand to create a seal, this prevents air from escaping out from the collar.

Use your free hand and feet to stroke and kick to the surface.

Gather and hold the blouse tightly at the collar and stomach level to prevent the blouse from losing air if it floats up too high.

Splash water on the blouse periodically to prevent the material from drying, dry material allows air to escape.

Repeat inflation as required.

Alternate Method
Turn the collar inside the blouse to help create a seal.

Unbutton the second button from the top.

Take a deep breath and bend forward slightly at the waist.

Place your mouth and nose inside the hole created by the open button and exhale one-half to three-quarters of a breath into the blouse.

Grasp material at the unbutton portion and pull downward.

Use your free hand and feet to stroke and kick to the surface.

Splash water on the blouse periodically to prevent the material from drying, dry material allows air to escape.

Repeat inflation as required.

Floating With Inflated Trousers

In warm water, trousers can be used as a primary expedient flotation device. However, in cold water, submerging your head to remove and inflate your trousers results in heat and energy losses that negate the benefit of using the trousers as a flotation device. Once your trousers are inflated, you float motionlessly as if wearing a life preserver. If needed, assume the heat escape lessening posture (known as the HELP position, see page 1-37) to slow heat loss. As trousers dry, air leaks out of the legs. To slow this process, occasionally splash water on the fabric. Reinflate trousers as needed.

Sling Method

The sling method works if you are a strong swimmer or naturally very buoyant. Take the following steps to inflate trousers using the sling method:

Take a deep breath, bend over, and remove your boots.

Note: Retain your boots. Tie the boot laces together and suspend the boots from your blouse or hang them around your neck so that they rest on your chest.

Remove your trousers. Button or zip the trouser's fly closed. This allows you to control airflow.

Tie the bottoms of the trouser legs in a square knot. (The appendix illustrates various knots.)

Ensure that the front (fly) of the trousers faces you.

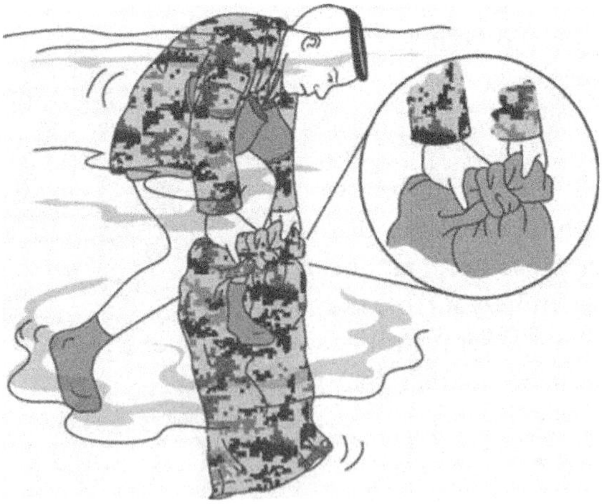

Hold the trousers above the water's surface and behind your head. Grasp both sides of the waistband and open with both hands.

Kick strongly to stay on top of the water while slinging the trousers overhead in order to trap air into them.

Once the waistband is submerged in the water, air is trapped in the legs.

Hold and seal the waistband underwater.

Slip the inflated legs over your head. Hold the waistband in toward your chest, the fly facing your body. To prevent air from escaping from the trousers, seal the waistband by either folding it or twisting it.

Lie back and relax, resting the back of your neck against the knot.

Splash water on the trousers periodically to prevent the material from drying. Dry material allows air to escape.

To replenish air in the trousers, you will use a technique known as the scooping method. With one hand on the open waistband, extend the trousers in front of you just below the surface of the water and scoop air bubbles with your free hand into the open waistband until the trousers have sufficient air. Repeat as necessary.

Splash Method

The splash method is an alternative to the sling method. As with the sling method, you must kick strongly to remain at the surface. To inflate trousers using the splash method, perform the following:

Take a deep breath, bend over, and remove your boots.

NOTE: Retain your boots. Tie the boot laces together and suspend the boots from your blouse or hang them around your neck so that they rest on your chest.

Remove your trousers. Button or zip the trouser fly closed. This allows you to control airflow.

Tie the bottoms of the trouser legs in a square knot. (The appendix illustrates various knots.)

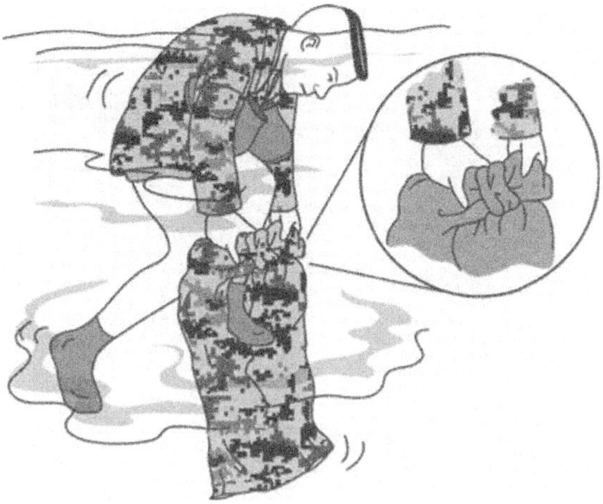

Ensure that the front (fly) of the trousers faces you.

Hold the trousers at the water's surface out in front of you by the waistband with the fly up.

Grasp the waistband at the surface with one hand. Insert your free hand into the waistband, palm down.

Flutter your hand rapidly to create bubbles. This sends a mixture of water and air bubbles into the trousers. The water passes through the fabric. The air remains trapped in the legs.

Hold and seal the waistband underwater.

Slip the inflated legs over your head. Hold the waistband in toward your chest, the fly facing your body. To prevent air from escaping from the trousers, seal the waistband by either folding it or twisting it.

Lie back and relax, resting the back of your neck against the knot.

Splash water on the trousers periodically to prevent the material from drying. Dry material allows air to escape.

To replenish air in the trousers, you will use a technique known as the scooping method. With one hand on the open waistband, extend the trousers in front of you just below the surface of the water and scoop air bubbles with your free hand into the open waistband until the trousers have sufficient air. Repeat as necessary.

Blow Method

The blow method is an alternative to the sling method. Use the blow method if you are a weak swimmer. Take the following steps to inflate trousers using the blow method:

Take a deep breath, bend over, and remove your boots.

NOTE: Retain your boots. Tie the boot laces together and suspend the boots from your blouse or hang them around your neck so that they rest on your chest.

Remove your trousers. Button or zip the trouser fly closed. This allows you to control airflow.

Tie the bottoms of the trouser legs in a square knot. (The appendix illustrates various knots.)

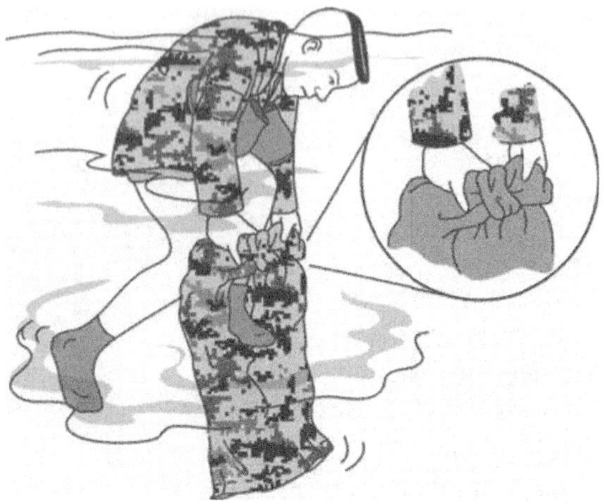

Ensure that the front (fly) of the trousers faces you.

Hold the trousers at the water's surface. Grasp both sides of the waistband and open with both hands.

Take a deep breath.

Drop 2 feet below the water's surface, pulling the waistband underwater.

Hold the waistband open with both hands and blow air into the trousers.

To fill the trousers with air, surface while keeping the waistband underwater, breathe in again, drop below the water's surface, and blow air into the trousers. Repeat these steps until the trousers are filled sufficiently. Once trousers are filled—

Hold the waistband underwater. Twist and pinch it off.

Slip the inflated legs over your head. Hold the waistband in toward your chest, the fly facing your body. To prevent air from escaping from the trousers, seal the waistband by either folding it or twisting it.

Lie back and relax, resting the back of your neck against the knot.

Splash water on the trousers periodically, to prevent the material from drying. Dry material allows air to escape.

To replenish air in the trousers, you will use a technique referred to as the scooping method. With one hand on the open waistband, extend the trousers in front of you just below the surface of the water and scoop air bubbles with your free hand into the open waist-band until the trousers have sufficient air. Repeat as necessary.

Avoiding Heat Loss in Cold Water

The rate of heat exchange in the water is about 25 times greater than it is in air of the same temperature. When you are immersed in cold water, hypothermia occurs rapidly due to the decreased insulating quality of wet clothing and as a result of water displacing the layer of still air that normally surrounds the body. You also lose about 50 percent of your body heat through your head; therefore, keep your head out of the water. Other areas of high heat loss are the neck, the armpits/sides, and the groin.

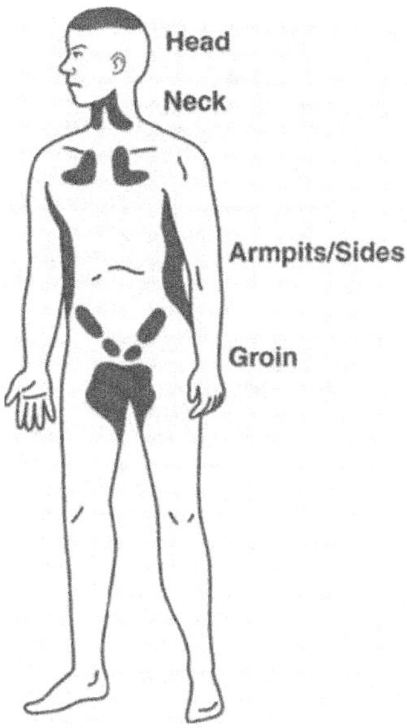

In cold water, DO NOT SWIM TO STAY WARM. Swimming, even with a slow and steady stroke, produces a lot of heat that is lost in the water. The heat loss can produce hypothermia that slows body functions and can result in serious injury or death. Remaining motionless conserves body heat three times longer than swimming. SWIM only if you have a flotation device and the shoreline is visible.

Individual Protection From the Cold

If you are equipped with a life preserver, assume the heat escape lessening posture (known as the HELP position) to slow heat loss and to protect major blood vessels near the body's surface. These areas lack insulating fat and are vulnerable to the chilling effects of cold water. To assume the HELP position—

Tuck your chin down tightly to cover your throat.

Draw your legs up in a fetal position to protect the groin.

Place your arms across your chest, tuck your hands into your armpits.

Wear some type of head covering (e.g., stowed cover, towel, handkerchief) to lessen heat loss through the scalp if head covering is available.

Group Protection From the Cold

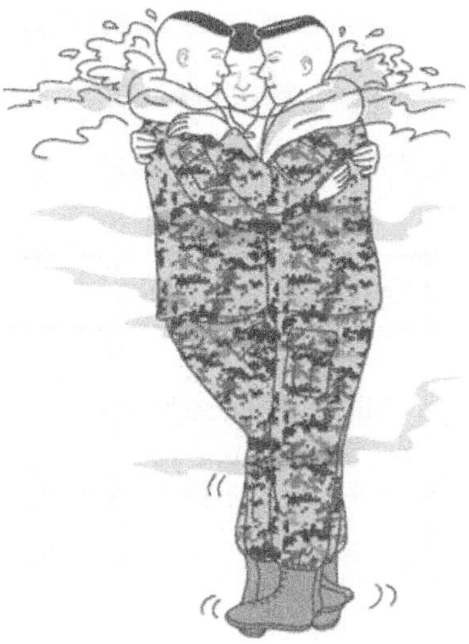

If three or more Marines are in the water and are equipped with life preservers, they should wedge tightly together and lock arms to form a circle known as a huddle position. This position protects vulnerable areas from heat loss.

A casualty who is suffering from the effects of the cold can be placed within the huddle to be surrounded by warmer water. If in the water for a prolonged period, it is recommended that everyone be rotated inside the huddle to maintain or re-warm each person's internal core temperature. If there are more than five Marines, they should make clusters of huddle positions.

Contact with other swimmers provides survival advantages—

Creates a larger target for search and rescue aircraft.

Provides additional warmth in cold water.

Improves morale.

Re-establishes the chain of command.

Reduces shock and panic.

Provides opportunities to administer first aid.

Supports exhausted Marines.

Drownproofing Methods

An object that floats on the surface has positive buoyancy. An object that floats a few feet beneath the surface has neutral buoyancy. An object that sinks has negative buoyancy. Most people have positive buoyancy and will float at the water's surface. To test your buoyancy—

Stand in water that is at your chest level or deeper.

Take a full breath.

Bend forward slowly.

Relax and wait.

If you have positive buoyancy, you will slowly rise to the surface. If you have neutral buoyancy, you will float a few feet beneath the surface. If you have negative buoyancy, you will sink. Regardless of whether you naturally float or sink, you can remain at the surface for extended periods without a life preserver if you exercise the appropriate drownproofing methods or survival strokes, which are based on your buoyancy.

Drownproofing methods consist of the T-method and the sweep. The breast stroke, side stroke, and elementary backstroke are the most common survival strokes. With any drownproofing method or survival stroke, remember the acronym **SAFE**:

SLOW EASY MOVEMENTS
Move slowly to conserve energy and minimize heat loss.

APPLY NATURAL BUOYANCY
Use natural buoyancy to support the body.

FULL LUNG INFLATION
Fill the lungs with each breath. Do not hold air in the cheeks.

EXTREME RELAXATION
Tight muscles are denser than relaxed ones and do not float as well.

Crawl Stroke

The crawl stroke, sometimes called the front crawl or free style, is the fastest stroke. To execute the crawl stroke—

Body position
Lie horizontal, on your stomach, in the water.

Look forward and downward at a 45 degree angle with the waterline between your eyebrows and hairline. Your head position is important as it assists in cutting a path through the

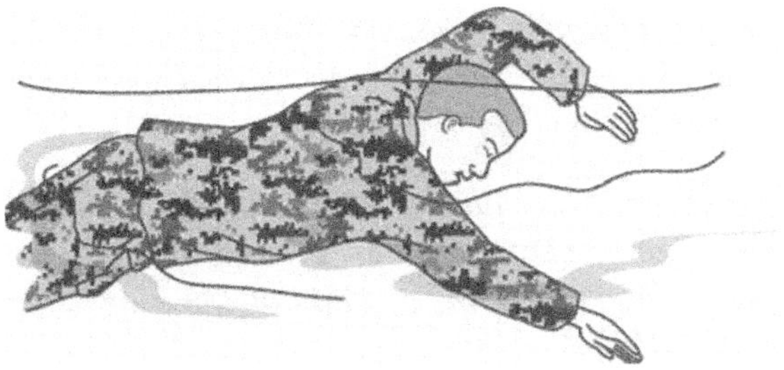

water. If your head is too high in the water, your lower body sinks significantly, making your stroke less efficient. If your head is too low in the water, water washes over your shoulders and neck, causing unnecessary drag.

Arm action

Arm action occurs in three phases: catch, propulsion, and recovery.

Fully extend one arm forward of your body, this positions your hand to catch the water in preparation for the propulsion phase.

To catch the water, bend your wrist (with your palm pointing outboard) and make an "S" shape (or inverted "S" shape) with your hand, ensuring that your hand does not cross the center of your body. Your left hand makes an "S" shape and your right hand makes an inverted "S" shape with your hand finishing at shoulder level.

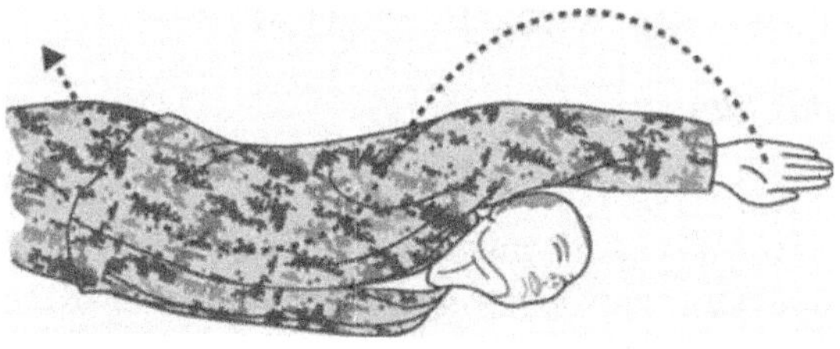

Push with your hands in a rearward fashion toward your feet until your arm is fully extended along your side, keeping your hands close to your body.

To begin the recovery phase, bend your arm at the elbow and raise your hand out of the water. Your hand breaks the surface of the water and maintains a height of 2 to 3 inches above the water's surface.

With your hand and arm moving in a forward manner, bring your hand past your head until your arm is about three-quarters of the way extended. At this point, turn your hand with your palm outboard so as to allow your thumb and fore-finger to enter the water first.

Once your hand enters the water, continue to push your arm forward until it is fully extended.

You are now prepared to catch the water again. These steps are performed in an alternating pattern: when one arm is catching and propelling, the other arm is recovering.

Leg action
Use a flutter kick to create the leg action for the crawl stroke. This kick is used for both propulsion and keeping the lower body horizontal with the water's surface. The flutter kick is an alternating leg action: one leg is kicking in a downward motion while the other leg is recovering to the surface to prepare for the next kick. The size of the kick ranges from 12 to 15 inches and depends on your height.

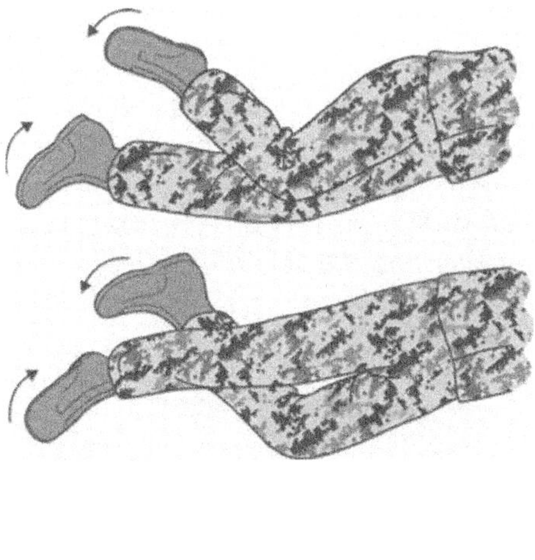

Maintain your legs in a semi-rigid manner.

Generate power for the kick from the hips.

Keep your feet loose at the ankles, they trail behind your legs and act as "flippers."

Execute the propulsion phase of this kick with a downward thrust of your leg.

Execute the recovery phase by pushing your leg back to the surface. This phase is complete when your foot reaches the surface.

Breathing
Breathe during the recovery phase of the arm action.

Roll your body and rotate your head to the side of your body where the arm recovery is occurring, this rolls the water away from your mouth. Keep your chin pushed back toward your shoulder.

Exhale while your face is still submerged and inhale when your face breaks the surface.

Breathe either bilaterally or rhythmically. To breathe rhythmically, breathe on the same side of your body every time your arm cycle occurs. Bilateral breathing requires you to breathe every one and a half arm cycles. To breathe bilaterally, breathe when your right arm is recovering, your face goes back in the water and the next time you breathe is when your left arm is starting its recovery three arm strokes later.

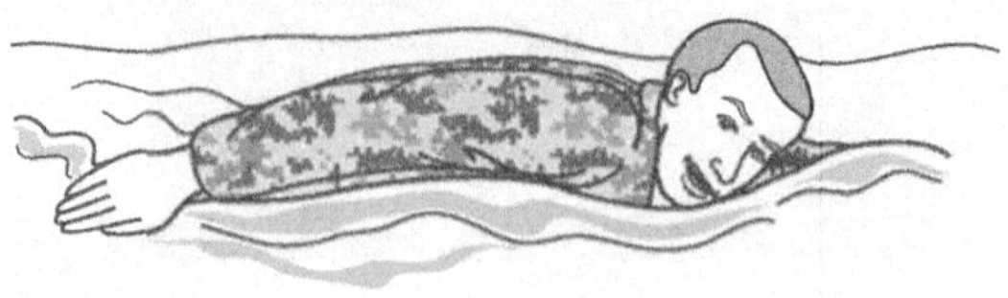

Bilateral breathing is the preferred method for breathing; it prevents the chance of hyperventilation and allows your body to maintain a lateral position in the direction you are swimming.

Coordination
This stroke uses constant arm and leg action.

T-Method

The T-method is a basic drownproofing method. This is the best survival technique if you have negative buoyancy. To execute the T-method—

With your face out of the water, take a deep breath and submerge your face in the water while holding your breath.

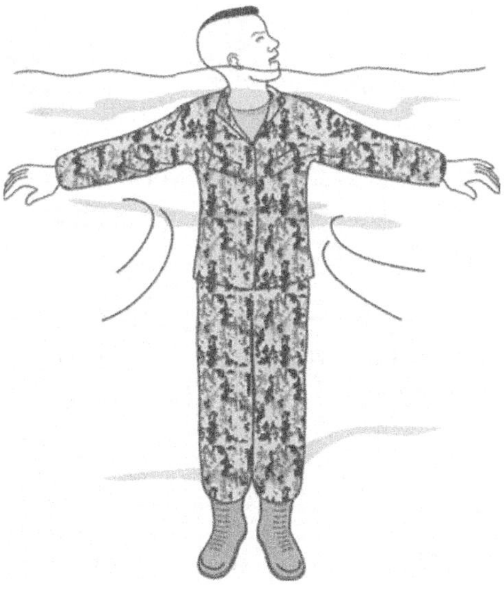

Float with your body in a horizontal position, arms extended from your side, and legs extended and joined.

Move your hands up to your armpits by tracing an imaginary line along your ribs.

Extend your arms outward (horizontal) to your sides, your body position resembles the letter "T."

Step out and forward with one leg and point your other leg to the rear, your knees should be slightly bent. Simultaneously, bring your arms down to your sides. Then bring your legs back together. You complete the steps by exhaling most of your air and preparing to surface your face to obtain another breath of air. Hold your head out of the water. Tilt your head back slightly. Breathe normally.

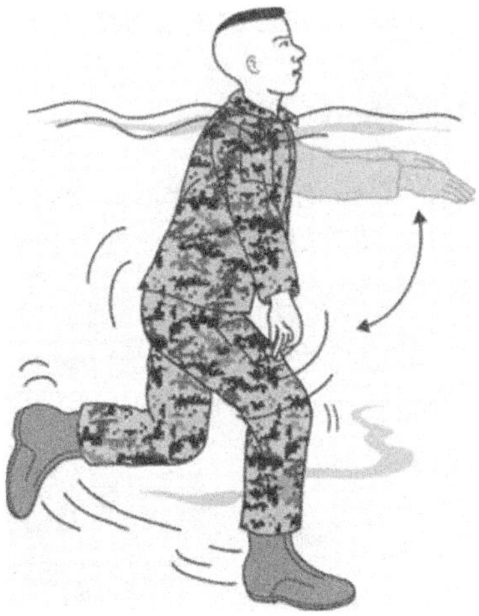

Once your breath is complete, move your hands up and down directly in front of your body. Do this two or three times to slow your descent into the water.

> NOTE: To avoid hyperventilating, hold your breath below the surface of the water for no more than 10 seconds.

The Sweep

The sweep works well if you have slight to excellent positive buoyancy. To execute the sweep—

Float face down in the water: bend 45 degrees at the waist, arms and legs dangling, head hanging down, relax all muscles.

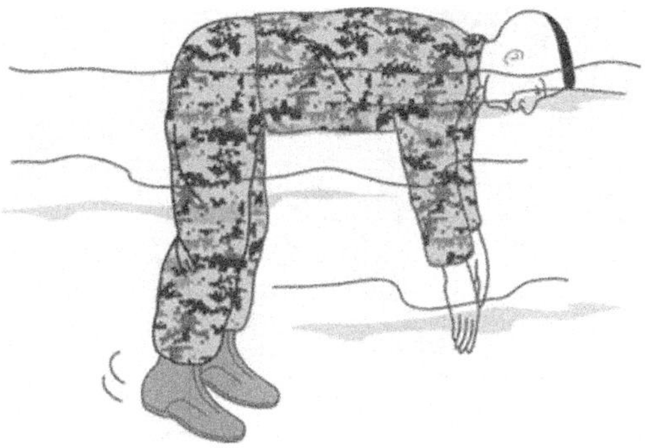

Spread your feet slowly to prepare for a single kick (one leg is forward and one leg is rearward).

Cross your arms in front of your chest, palms outboard with the back of each hand touching the opposite ear.

Exhale prior to raising your head for a breath.

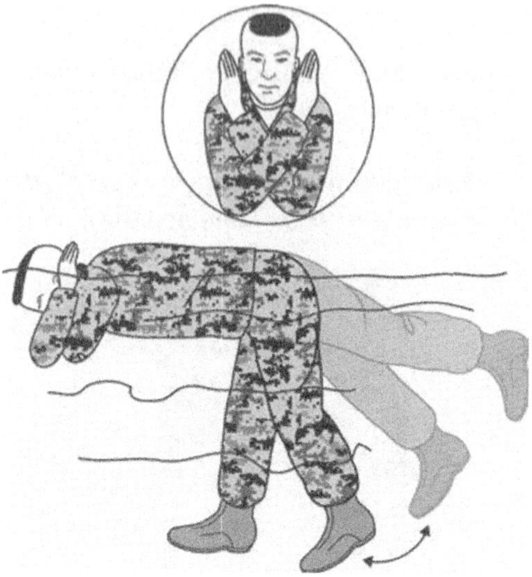

Bring your legs together and sweep your arms down and out until the arms are fully extended out to the sides. This raises your face above the water and allows you to catch a breath of air.

With your air supply replenished, return your face to the water and relax while sweeping your arms in a downward motion in front of your body to prevent/slow your descent. Do not hold your breath for more than 10 seconds. More than 10 seconds enhances your chances of shallow water black out and subsequent drowning.

Breast Stroke

Use this stroke to swim underwater, through oil or debris, and in rough seas. If you are a good swimmer and not wearing combat gear, the breast stroke is the best stroke for long-range swimming because it provides good visibility and allows you to conserve your energy and maintain a reasonable speed.

Body position
Lie prone in the water. Swim with your trunk and legs projecting back and down at an angle of 20 to 30 degrees.

Extend arms out in front (hands together [side by side]), and extend legs behind (toes pointed) to prevent drag.

Face downward, looking forward at a 45 degree angle to break the water and to prevent water from washing into the collar area causing drag.

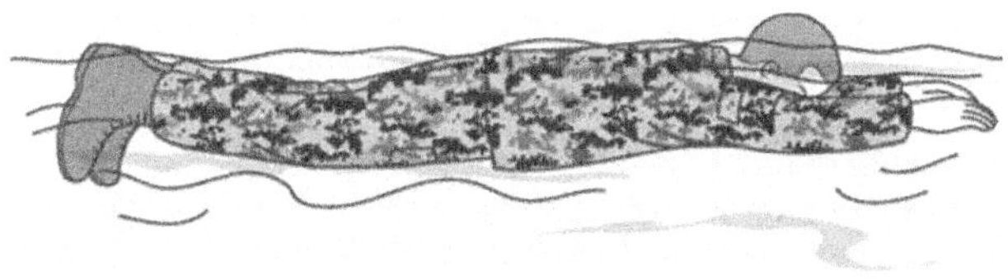

This is known as a glide.

Arm action

Turn your palms outward and bend your arms slightly.

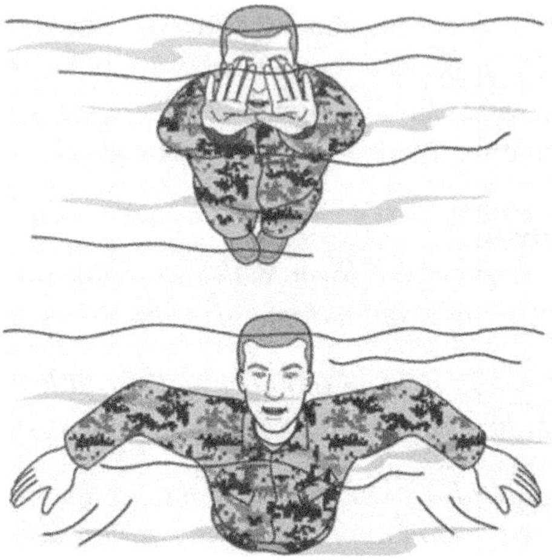

Sweep your arms sideward and slightly downward until your hands are opposite and slightly below your shoulders.

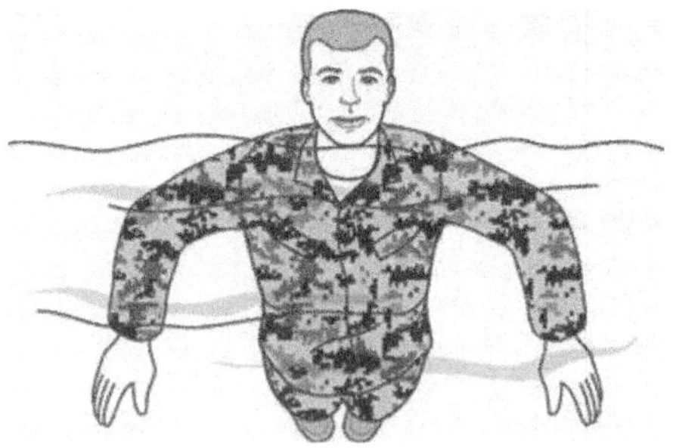

Rotate your head up, breathe once your mouth breaks the surface.

Bring your hands and arms up along your chest and thrust them forward until they are extended and ready to execute the next arm pull.

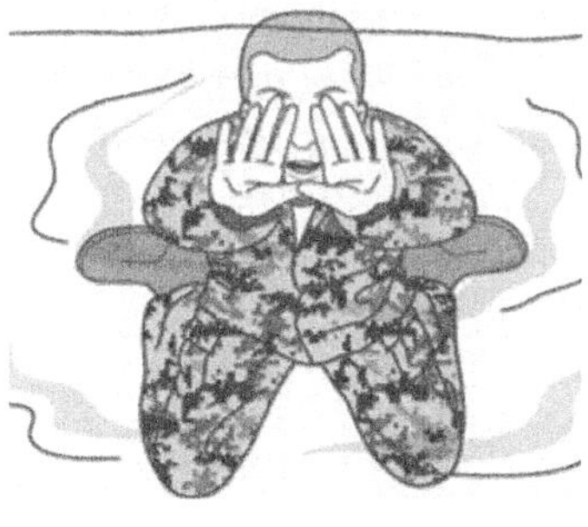

As the arms start their recovery into the glide, the head should rotate forward, resubmerging the face.

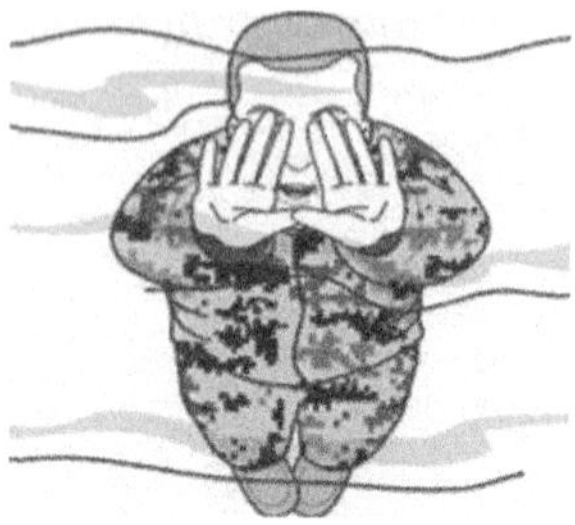

Leg action

Draw your heels toward your buttocks, establish a 45 degree bend in the knees.

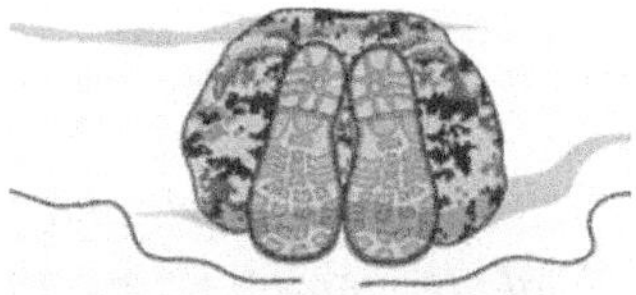

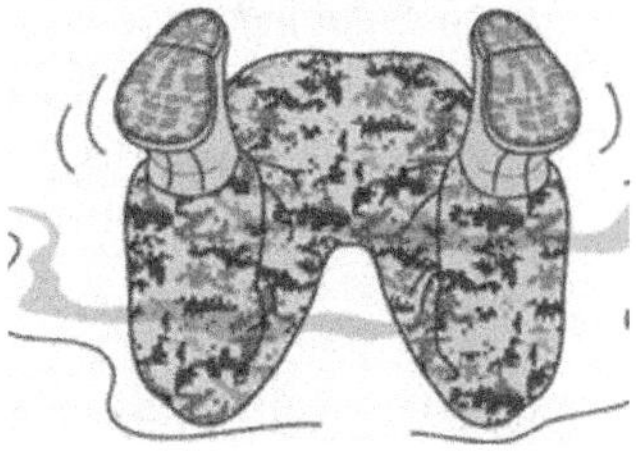

Thrust your legs outward and rearward, then squeeze them together. The whipping action of the feet aids forward propulsion.

This is known as the breast stroke kick.

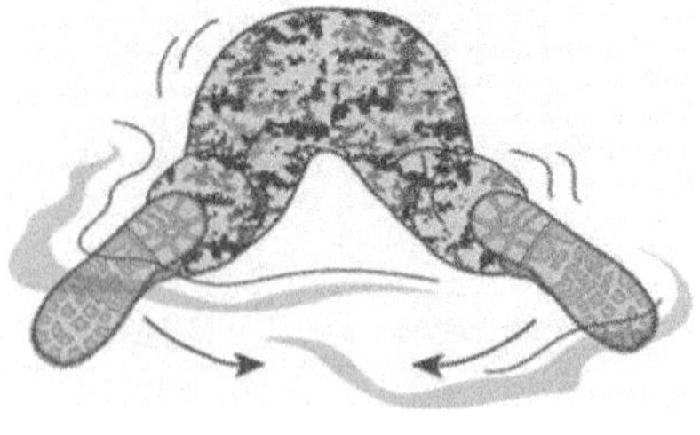

Breathing
Inhale during the arm pull and exhale through your mouth and nose during the finish of the breast stroke kick and glide.

Coordination
The stroke movement is in three counts:

Begin your arm pull. Near the finish of the pull, flex your knees and bring your heels toward your buttocks. The arm pull counteracts the resistance created by the knees.

As the arm pull is completed, thrust your hands forward, kick your legs outward and rearward, and squeeze them together.

Glide through the water for approximately 1 to 3 seconds or until your forward momentum decreases, then begin the next stroke.

Side Stroke

The side stroke is a survival stroke because you use both arms for buoyancy, with each arm creating a slight propulsion. The majority of your body's propulsion comes from your kick. To execute the side stroke—

Body position
Lie on your side with your lead (bottom) arm extended beyond (with a slight bend in your elbow) your head and in line with your body. Palm is down and your hand is submerged 6 to 8 inches.

Extend your trail (top) arm down the length of your body over your thigh.

Keep your legs straight and together, toes pointed rearward.

Keep your face out of the water, this allows for free breathing.

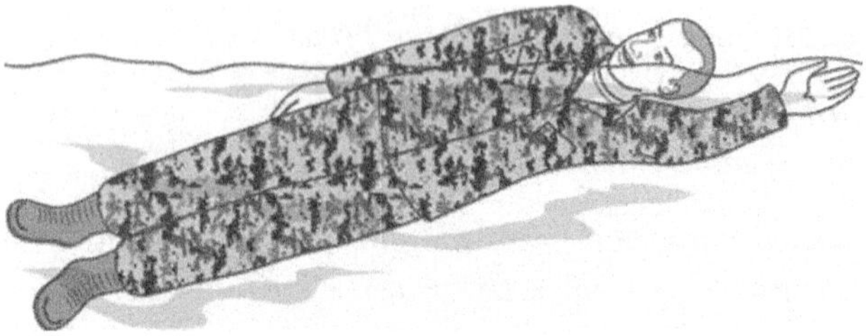

This is known as the glide.

Arm action

With your lead arm, pull your arm downward, while flexing at the elbow, until it is straight down from your shoulder.

Rotate your shoulder and pull your elbow into your side. This should put your lead hand at shoulder level. At the same time, turn your palm toward your face and thrust forward to your original, extended position.

Draw your right hand upward in front of your chest to shoulder level. Rotate your palm toward your feet, then push it downward in front of your body toward your feet to catch the water.

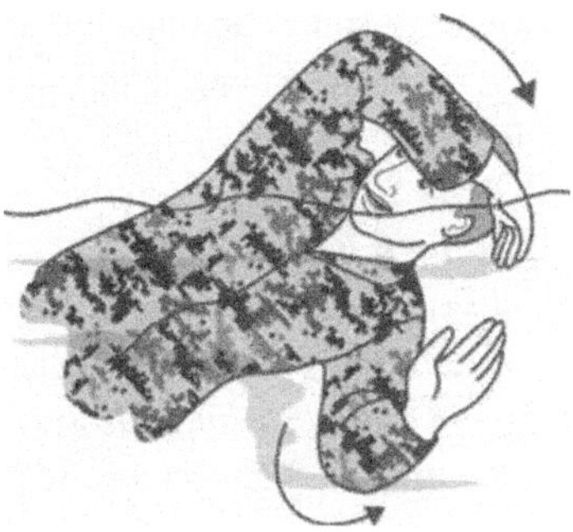

Push your trail hand backward to its original position on top of your thigh. (Your trail hand starts forward and meets your lead hand at your chest/shoulder.)

Leg action

To perform the scissor kick, the top leg always goes forward and the bottom leg always goes rearward. From the extended position, draw—or recover—your feet toward your buttocks until your legs are bent at a 45 degree angle at the knees and the hips are flexed at a 45 degree angle with the thighs.

Once the legs have completed their recovery and while maintaining a 45 degree bend in the knees, extend the legs fully into a "V" shape in order to catch the water for the propulsion phase.

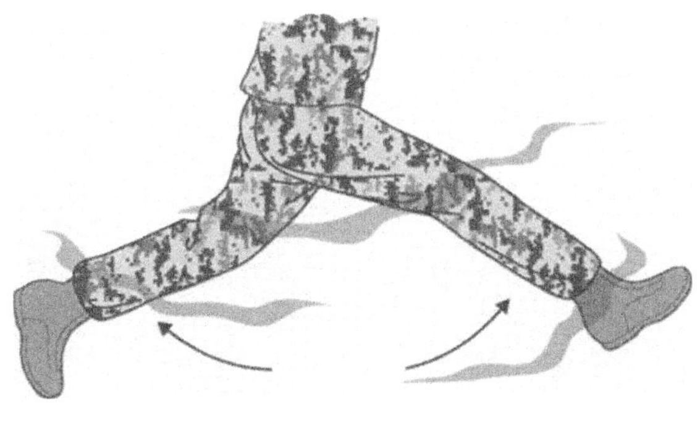

Once the legs are separated and extended forward and rearward to the "V" position, sweep the legs together until the feet are together.

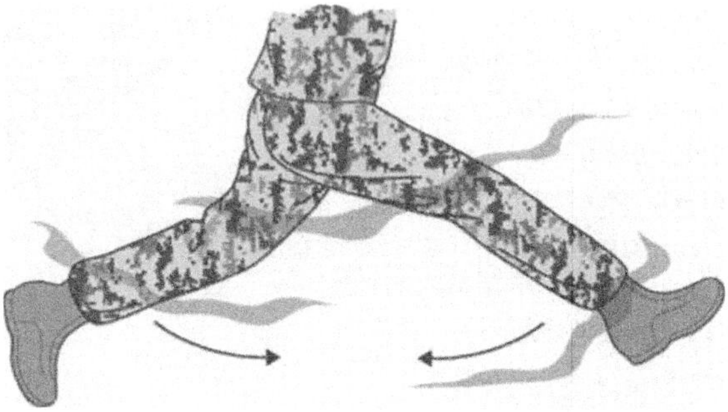

You are now in the glide position.

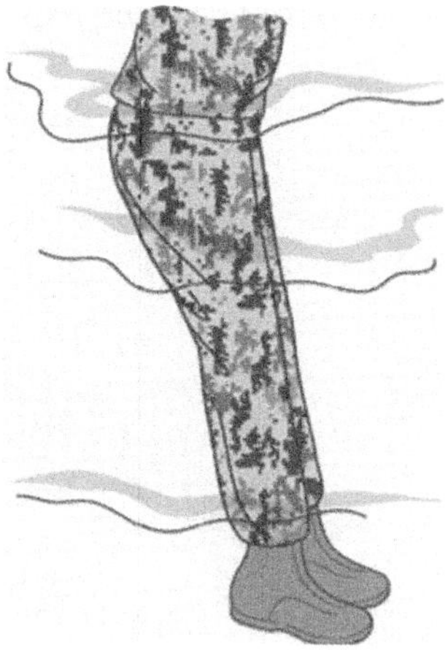

Breathing

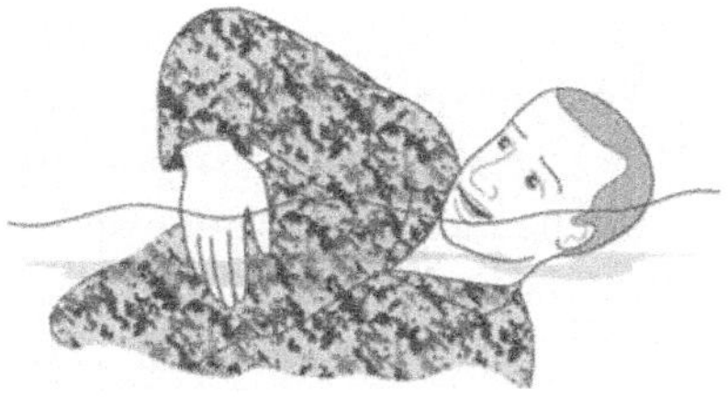

As long as your face remains clear of the water, it is a free breathing stroke. However, it is recommended that you exhale then quickly inhale when the legs are sweeping back together in the scissor kick. This is when the body reaches its highest point in the water, thus clearing the face completely from the surface of the water making it the optimum time to breathe.

Coordination

Begin the stroke with the downward pull of your lead arm. At the same time, bring your trail arm upward and draw your knees up to begin the kick. Let the thrust of the lead arm, push of your trail arm, and the kick of your legs coincide in order to finish the glide position. Glide through the water for approximately 1 to 3 seconds or until your forward momentum decreases, then begin the next stroke.

Elementary Backstroke

The elementary backstroke is also an excellent survival stroke. It relieves the muscles that you use for other strokes, and it is the recommended stroke for weak swimmers or nonswimmers. To execute the elementary backstroke—

Body position
Start on your back.

Face up, chest up, and hips up, keeping an arch in your lower back with arms pressed to your sides and your legs extended and joined to prevent drag.

Arm action
Trace your hands up your sides to an area near your armpits then extend your arms out to the sides to form the letter "T" (palms facing feet), locking out the elbows.

> NOTE: Don't raise your arms above your head. This creates drag, changes your body position, and submerges the head.

Slap your palms to your thighs using a strong sweeping motion.

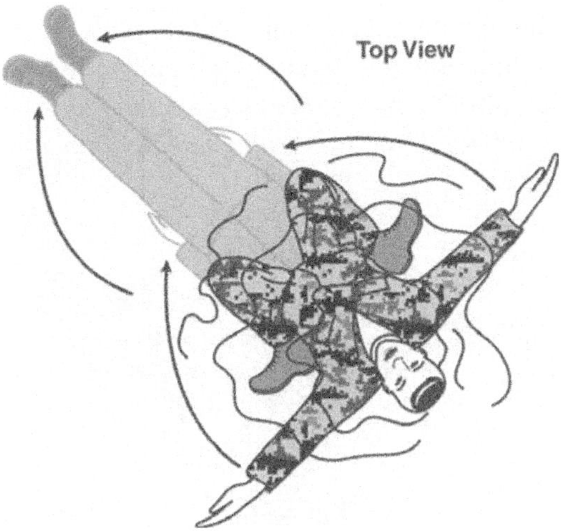

Leg action
Bend both legs at the knee (90 degree angle) slightly separating your knees and drawing your heels downward to a point under and outside your knees. The knees are spread as wide as the hips or slightly wider depending on the body type of the swimmer.

Circle around in a whipping action, ending with legs in a glide position.

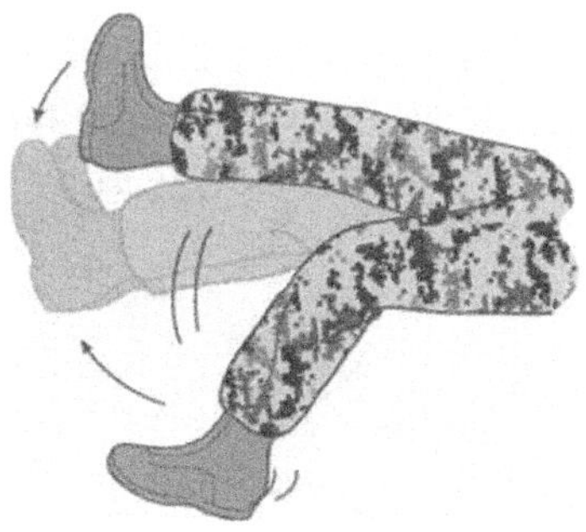

Breathing
Breathe anytime during this stroke. However, it is recommended that you exhale then quickly inhale when your arms are sweeping back toward your sides and while your legs are sweeping back together. This is when the body reaches its highest point in the water, thus clearing the face completely from the surface of the water and making it the optimum time to breathe.

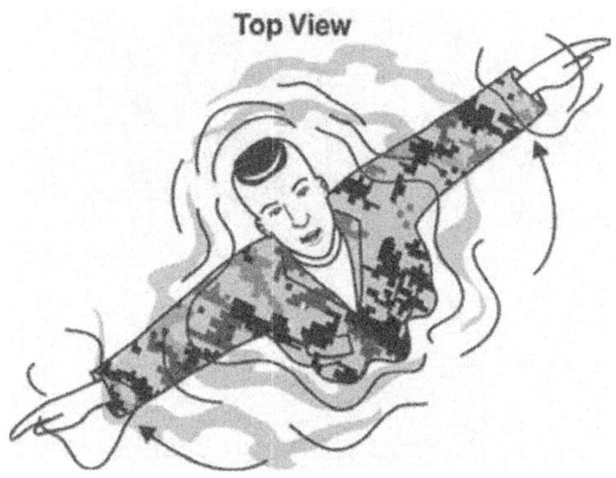

Coordination
The stroke movement occurs in three counts (recovery, catch, power).

Begin the arm pull (recovery).

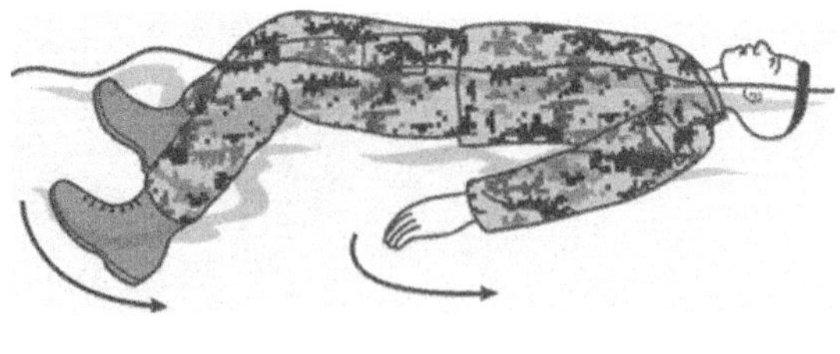

Near the finish of the pull, flex your knees to a 90 degree angle. The arm pull counteracts the resistance created by the knees.

Kick out your legs, and squeeze them together as the arm pull is completed (catch, power).

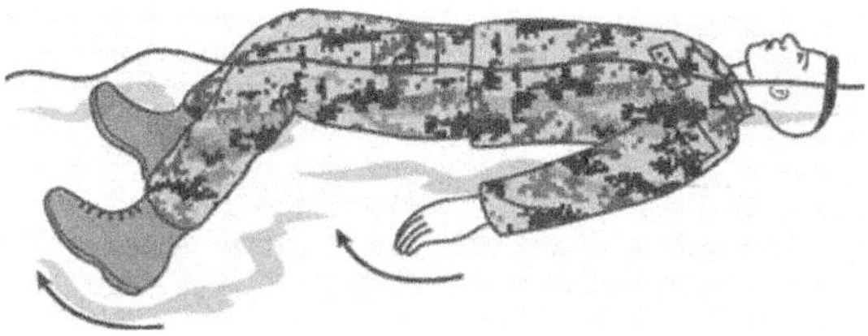

Glide through the water for 1 to 3 seconds or until your forward momentum decreases, then begin the next stroke as your momentum slows.

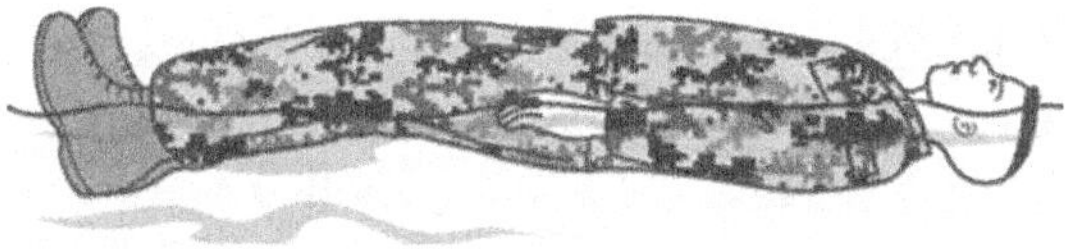

Chapter 2
Water Rescues

A drowning victim can panic and react with unexpected violence and can seize and inadvertently drown a rescuer. Therefore, if possible, a water rescue should be executed from a distance. Reaching, wading, or throwing methods are used in a distance rescue.

If a victim is too far away to use these methods, a swimming rescue may be necessary. Remove all combat gear, when possible, before entering the water. Swim within 2 to 6 yards of the victim to maintain a margin of safety, this allows you to reassess the situation and reassure the victim. If the victim is unconscious, use the wrist tow method or cross-chest carry method to pull the person to safety. If the victim is struggling, use a rear approach and then execute either a single armpit level off or a double armpit level off before towing the victim to safety.

If the victim does begin to overpower you, there are techniques that allow you to defend yourself without having to abandon the rescue. These techniques include the block, the wrist-grip escape, the front head-hold escape, and the rear head-hold escape. These techniques allow you to separate yourself from the victim, reassess the situation, and then attempt the rescue again.

> *Note:* In this chapter's illustrations, the rescuer is shown without a helmet.

Reaching Rescue Techniques

Reach

From a safe position at the water's edge, reach out to the victim. Talk constantly to calm the victim. Retain partial contact with land or some solid support structure (e.g., pier, bridge). If the victim is close but still beyond reach, extend an object (e.g., stick; pack; rifle with magazine removed, chamber empty, and muzzle pointing toward victim) that the victim can grasp. Pull the victim slowly to safety. Once the victim is close to shore, remove the victim from the water.

Note: You can also extend a foot to the victim if you can retain a secure grip on a solid support structure.

Reach From a Deck

Reach from a deck rescue can be executed by either a swimmer or a nonswimmer, and it can be used on an active or a passive victim. Be sure to reassure the victim during the rescue. To execute reach from a deck—

Lie prone on the deck with your body firmly anchored. To anchor your body, lie flat, spread legs apart, and extend one arm behind you with your palm down and on the deck.

Keep as much of your weight on the deck as possible and extend your free hand to the victim.

Grasp the victim's wrist from above, your thumb and index finger are facing you. The victim should never be allowed to grab you and put your life at risk. Therefore, never reach across the victim to grab his wrist, always grab the wrist that is the closest to you.

Keep your arm straight and locked out and pull the victim to the side of the deck. The victim should never be allowed to grab you and put your life at risk. Therefore, never pull the victim into you, always pull the victim into the side.

Arm Extension

An arm extension rescue can be used if you cannot reach the victim using the reach from a deck rescue technique and you must enter the water. An arm extension rescue technique is used for a victim who is either active or passive. Once you determine that a reach from a deck rescue technique is not viable—

Reassure the victim and quickly ease into the water while holding onto the deck with one hand.

Grasp the victim's wrist from above, your thumb and index finger are facing you. The victim should never be allowed to grab you and put your life at risk. Therefore, never reach across the victim to grab his wrist, always grab the wrist that is the closest to you.

Keep your arm straight and locked out and pull the victim to your side. The victim should never be allowed to grab you and put your life at risk. Therefore, never pull the victim into you, always pull the victim into the side.

Leg Extension

If the victim is beyond the reach of your arm, ensure that you have a firm grip on the deck, extend a leg to the victim and allow him to grab it, slowly bring the victim in closer until you can grab the victim's wrist that is holding onto your leg. Once you have a firm grasp on the victim's wrist, use the steps in the arm extension to pull the victim to safety. The leg extension rescue can only be used to rescue an active victim because the victim must be able to grab the rescuer's leg.

Wading Assist

Do not wade into water that is deeper than your chest. Talk constantly to calm the victim. If possible, do not touch the victim directly. Extend an object (e.g., stick; pack; rifle with magazine removed, chamber empty, and muzzle pointing toward victim) that the victim can grasp. Once the victim grasps the object, pull the victim slowly to safety.

Throw

If the victim is not within reach, use an expedient line to throw a lifesaving device to the victim. A lifesaving device can be any weighted item that floats (e.g., a canteen that is one fourth full of water). The lifesaving device must be secured to the end of the rope so that the rope will feed out from its coil when tossed. Talk constantly to calm the victim. Once the victim grasps the line or the lifesaving device, pull the victim toward you at a steady pace that keeps the victim's head above the water's surface. DO NOT pull so strongly that you break the victim's grip on the line. To prepare and use an expedient line and lifesaving device—

Tie a bowline at one end of the rope. (The appendix illustrates various types of knots.)

Unfasten the lid of a canteen.

Place the bowline around the neck of the canteen.

Refasten the lid so the canteen hangs from the bowline loop.

Place one end of the rope under the ball of your forward foot to secure it, either tie a knot at the end of the rope or tie an object at the end of the rope to create a block. Stand with your weight on the end of the rope. DO NOT tie the rope around your ankle.

Coil 20 to 30 yards of the rope, and hold it in a nonthrowing hand.

Place the canteen in your throwing hand.

Use an underhand throw to pitch the canteen and rope a short distance over the victim's head. Keep your nonthrowing hand open so the coil can unfold freely. The rope should trail across the victim's outstretched arms.

Retrieve the rope if the throw is inaccurate or the victim fails to grasp it. Recoil the rope as it is retrieved.

Divide the coil and throw again.

Lifesaving Approaches

Properly approaching the victim is as important as any other aspect of a rescue. Knowing the appropriate and safest method needed to approach a victim is critical in ensuring your safety and the victim's survival. Determining the victim's physical state (distressed swimmer, victim that is active and drowning, or victim that is passive and drowning) is crucial and will determine which type of approach you will execute.

Front Surface Approach

The front surface approach is typically performed when the victim is passive. You must remember that approaching the victim from the front is dangerous because a distressed swimmer or victim that is active can lunge toward you.

Rear Approach

If the victim is active, you must remember that his response can change rapidly. Therefore, approaching him from the front could be extremely dangerous. The victim could easily grab you and take you under the water during a state of panic. Therefore, use a rear approach when possible. When approaching a victim from the rear, reassure the victim until you get to within 2 to 6 yards of the victim; at that point, stop talking to the victim in order to maintain an element of surprise.

Approach Strokes

Before approaching a victim, you must evaluate the victim:

- Is the victim a distressed swimmer?
- Is the victim active and drowning?
- Is the victim passive and drowning?

Select your approach stroke based on your evaluation. Remember, the safety of the rescuer is paramount. Do not endanger yourself in an attempt to reach a drowning victim.

Crawl Stroke Approach Stroke

The crawl stroke (free style) is the fastest approach stroke. This stroke is used when the victim is passive and/or unconscious in the water. That is, the victim is face down, submerged or near the surface of the water, is not breathing, or is not moving. If any of these conditions exist, there is an obvious need to reach the victim as quickly as possible.

Execution of the crawl stroke approach stroke is the same as the crawl stroke (see pages 1-40 through 1-44), except that your head remains above the water's surface to allow for free breathing and maintaining eye contact with the victim.

Maintaining eye contact is critical. If the victim becomes submerged, you will have a better chance of locating him if you know his last location. Your distance from the victim will determine if your head remains out of the water the entire time you are executing the stroke. If you are 54 yards or less from the victim, you should maintain eye contact by keeping your head raised until you reach the victim. If you must swim more than 54 yards to reach the victim, make eye contact with the victim, place your face back in the water for four or five strokes, then raise your head again to regain eye contact with the victim (repeat these steps until you reach the victim).

Body position
The position of your body is horizontal to diagonal and influenced by the position of the head. This allows you to maintain eye contact with the victim.

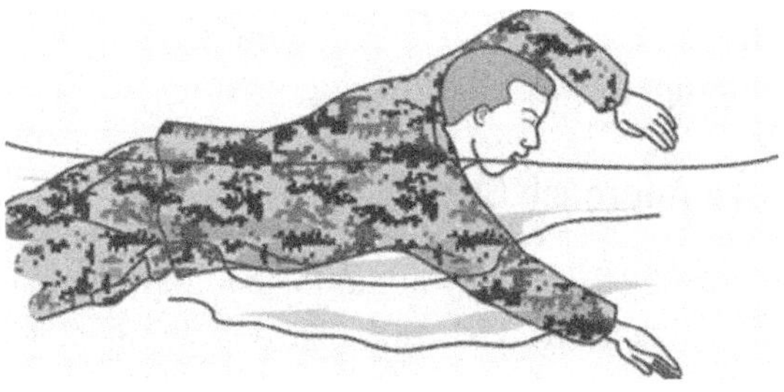

Arm action
While one arm is forward of your head to catch the water for the propulsion phase, extend your other arm alongside your body in what is known as the recovery phase. The recovery phase ends when your arm is extended forward of your head to begin the propulsion phase. In the propulsion phase, your arms are approximately shoulder-width apart and the pulling action is slightly wider and deeper to compensate for your raised head.

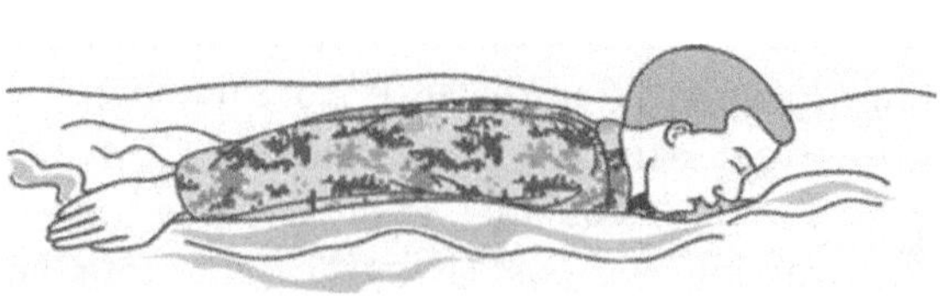

Leg action

Kick one leg in a downward motion while your other leg is recovering to the surface to prepare for the next kick (this is known as a flutter kick). This kick is used for propulsion and maintaining the lower body as horizontal to the water's surface as possible. When your head is raised, your knees are slightly bent in order to keep your feet near the surface.

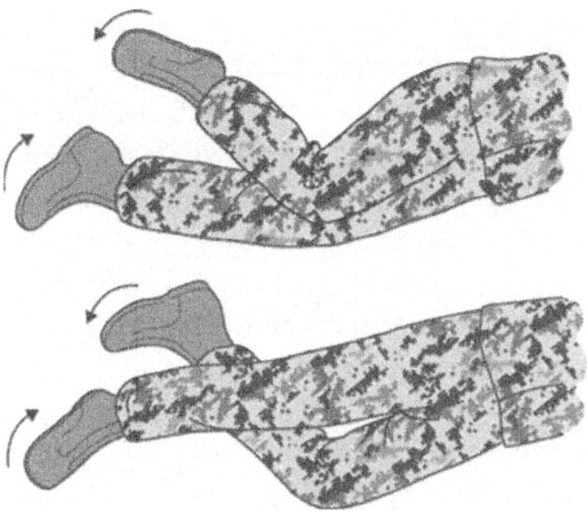

Breathing

If your face is above the water, breathe as needed (free breathing).

If traveling in excess of 54 yards to reach the victim, rotate your head to the right or left during the recovery phase and inhale when your face is out of the water or wait until your head is out of the water and facing forward while regaining sight of the victim.

Coordination

This stroke uses constant arm and leg action.

Breast Stroke Approach Stroke

The breast stroke approach stroke is used when a swimmer is distressed or a victim is active and drowning. This stroke is also used when the victim has a suspected spinal injury because it minimizes the wake created around the victim, minimizes the movement of the victim's head and neck, and helps prevent further injury.

If a victim is at a distance greater than 54 yards and if the swimmer is in distress or the victim is active and drowning in open water (ocean, lake), you should use the crawl stroke approach stroke to rapidly approach the victim, stopping approximately 11 yards from the victim in order to assess the situation.

Execution of the breast stroke approach stroke is the same as the breast stroke with a few minor deviations.

Body position
The position of your body is horizontal to diagonal and influenced by the position of the head. This allows you to maintain eye contact with the victim. It also allows you to communicate with the victim in order reassure him as you approach.

Arm action

Execute the arm action for the breast stroke, except that your arm pull is wider and deeper, which allows your head to remain above the water's surface.

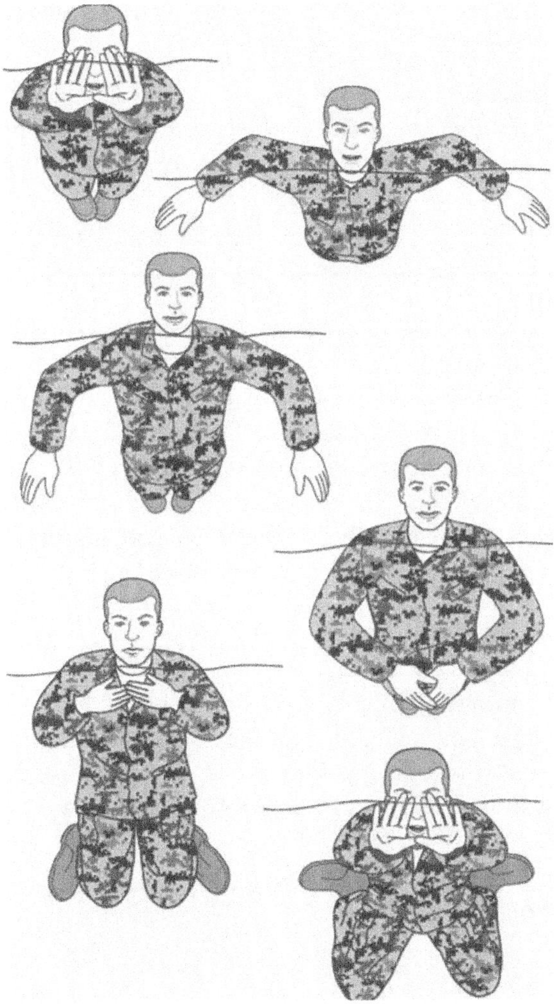

Leg action
Execute the leg action for the breast stroke, except that your legs open wider during the propulsion phase of the stroke.

Breathing
Keep your face above the water and breathe as needed (free breathing).

Coordination
Coordination is the same as the breast stroke, except that there is no glide period. You constantly stroke until you reach the victim.

Level Offs

A properly performed level off positions the victim's face up and horizontal to the water's surface. There are three types of level offs: front surface approach, single armpit level off, and the double armpit level off. The single armpit level off and the double armpit level off are done by performing a rear approach, which is the preferred approach when a victim is active.

Front Surface Approach

The front surface approach is performed when the victim is passive. Approaching a victim from the front may place you in danger because a distressed swimmer or a victim who is active may lunge toward you. To execute a front surface approach—

Stop 2 to 4 yards from the victim to reassess the situation and reassure the victim.

Determine which arm to use in order to rotate the victim into a face-up position (your right hand on the victim's right wrist or your left hand on the victim's left wrist).

Move into position and prepare to grasp the victim.

Turn sideways and move toward the front of victim.

Once in position, reach forward and grab the victim's wrist. Your thumb is on the inside of the victim's wrist, as if you were checking his pulse with your thumb. Your remaining fingers wrap around the victim's wrist.

Lean back immediately and execute a powerful scissor kick or inverted scissor kick and perform short, vigorous pulls with your free arm.

As the victim begins to move forward in the water, kick, pull, and twist outboard on the victim's wrist. The momentum created from the kick and the pulling and twisting action of the victim's wrist will rotate the victim into a face-up and horizontal position in the water.

Extend and lock out your towing arm down the length of your body and execute a wrist tow to move the victim to safety.

Single Armpit Level Off

Approach the victim.

Stop 2 to 6 yards from the victim to reassess the situation, reassure the victim, and maintain your margin of safety.

Approach the victim slowly and grab his armpit with your hand (your right hand to victim's right armpit or your left hand to victim's left armpit). Position yourself sideways to the victim and place your elbow in the center of his back.

Pull with your hand that is in the victim's armpit while pushing with your elbow that is in the victim's back to place the victim horizontal on the water's surface (face up). To assist in placing the victim horizontally, use your free arm to execute short, vigorous pulls and use your legs to execute a scissor kick or inverted scissor kick.

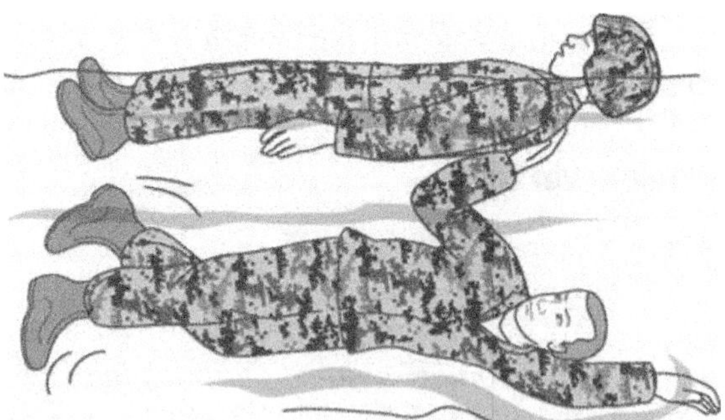

Once the victim is horizontal and on the water's surface, begin forward momentum by extending your towing arm to a fully locked-out position and executing a single armpit tow.

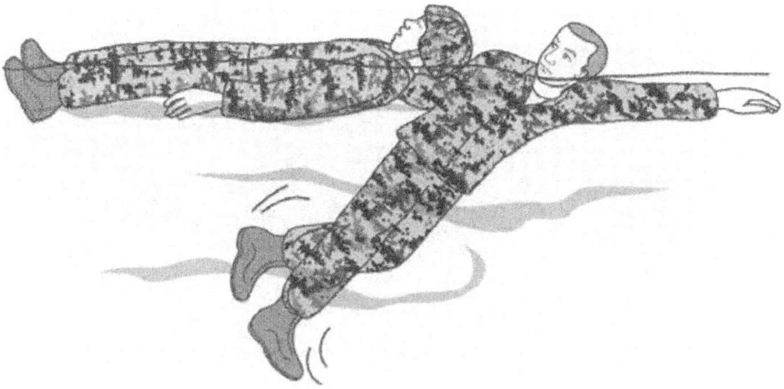

Double Armpit Level Off

Approach the victim.

Stop 2 to 6 yards from the victim to reassess the situation, reassure the victim, and maintain your margin of safety.

Approach the victim slowly and grab his armpits with both your hands. Place your elbows on his back.

Pull with your hands that are in the victim's armpits while pushing with your elbows that are in the victim's back and use an inverted breast stroke kick to place the victim horizontal on the water's surface (face up).

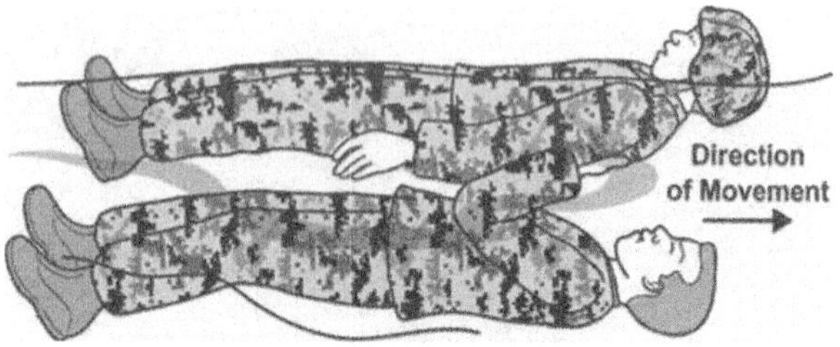

Once the victim is horizontal and on the water's surface, begin rearward momentum by slowly extending your arms into a fully locked-out position and executing a double armpit tow and inverted breast stroke kick.

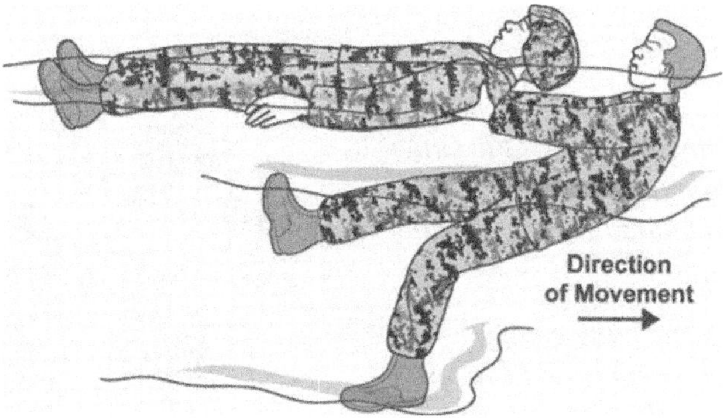

Rescue Techniques

Wrist Tow

Use the wrist tow method to rescue a victim who is floating face down. DO NOT use the wrist tow on a struggling victim. If time allows, remove your helmet and gear before attempting the rescue. Swim toward the victim using a modified breast stroke. Swim within 2 to 6 yards of the victim to maintain a margin of safety, this allows you to reassess the situation and reassure the victim. The following steps show proper front surface approach, wrist tow procedures:

Approach the victim from the front and grasp the underside of the victim's left wrist with your left hand or the right wrist with your right hand. Ensure that your thumb is on the underside of victim's wrist.

Lean back, pulling and kicking strongly to move the victim into a horizontal position.

Twist the victim's wrist to rotate the victim into a face-up position.

Swim toward safety using the lifesaving stroke.

> NOTE: The lifesaving stroke is a modified side stroke wherein the top arm is used to tow or carry a victim to safety and use a scissor kick or an inverted scissor kick is used for propulsion.

Keep a firm grip on the victim's wrist.

Keep your towing arm fully extended and along your side. This keeps the victim in the water column and prevents drag.

Ensure the victim's head does not go under water during the recovery.

Single Armpit Tow

You perform the single armpit tow after a victim has been properly leveled off. The single armpit tow uses the lifesaving stroke. To execute the single armpit tow—

Place one hand (top arm) in the victim's armpit. Your towing arm is straight and along your side.

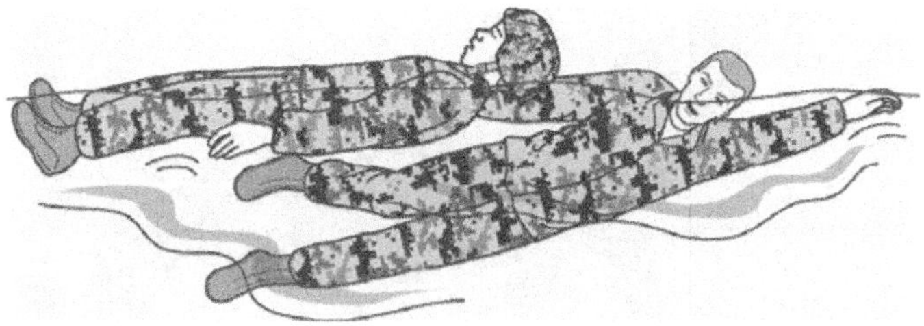

Use your lead arm (bottom arm) to execute short, vigorous pulls.

Execute a scissor or inverted scissor kick in a continuous and vigorous manner.

Use either free breathing or explosive breathing during the tow. During free breathing, keep your head above the water's surface and continuously reassure the victim. During

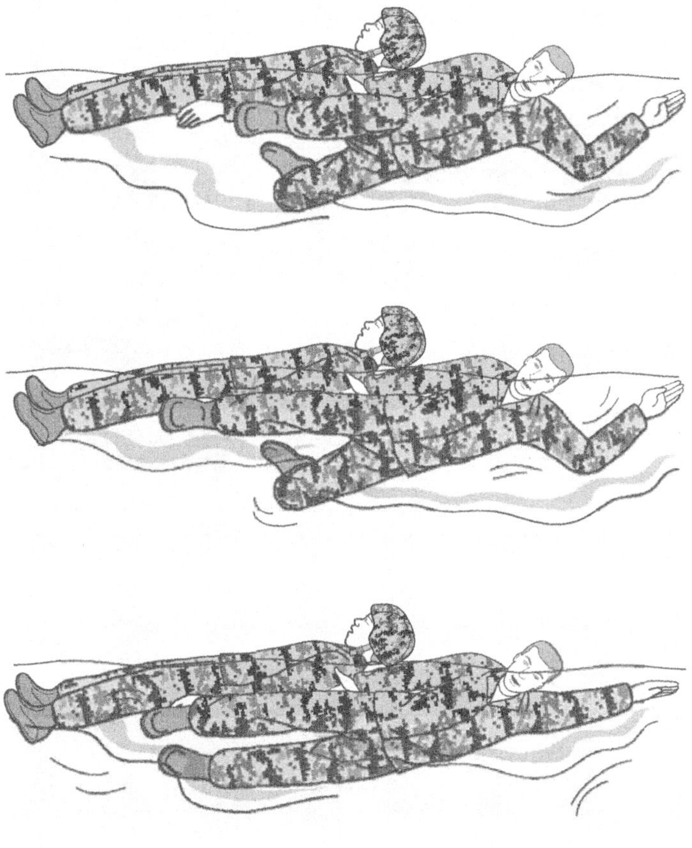

explosive breathing, put your head under the water to plane off your body's angle and reduce drag. Execute approximately two strokes, then lift your head up to breath and reassure the victim. Return your head back underwater for the next few strokes.

Double Armpit Tow

You perform the double armpit tow after a victim has been properly leveled off. To execute the double armpit tow—

Place both hands in the victim's armpits.

Extend both arms fully along your body. You are on your back, and your face is clear of the water.

Use free breathing.

Reassure the victim at all times.

Use the inverted breast stroke kick (it is the only kick that can be used due to your body position) to tow the victim to safety. (The inverted breast stroke kick is the same kick used in the elementary backstroke.) Your kick must be continuous and vigorous in order to keep the victim's face above the water.

Collar Tow

You perform the collar tow after a victim has been properly leveled off. To perform a collar tow using the victim's blouse—

Maintain control of the victim by grasping his armpit with one hand and then with your free hand grasp either his combat gear or his blouse between his shoulder blades. If grasping the blouse, grasp the material with your palm up, then turn your hand over to tighten the material.

Release the victim's armpit once control is established and execute the lifesaving stroke to tow the him to safety.

Cross-Chest Carry

Use the cross-chest carry to carry a victim to safety if the victim is struggling or when moving through heavy surf. Remove your helmet and gear before attempting a rescue. Talk constantly to calm the victim. Swim toward the victim using a breast stroke approach stroke. Swim within 2 to 6 yards of the victim to maintain a margin of safety, this allows you to reassess the situation and reassure the victim. The following steps illustrate proper cross-chest carry procedures:

> **CAUTION:** The cross-chest carry causes fatigue even if you are in excellent physical condition.

Use a level-off technique to place the victim in a horizontal, face-up position.

Retain a grip on the victim with one hand. Reach over the victim with your free hand to encircle the victim's chest. Place your free hand on the victim's opposite rib cage, just below his armpit.

Release your grip once you have a secure hold on the victim's chest.

Swim toward safety using the lifesaving stroke while keeping a firm grip on the victim's chest and your hip on his back.

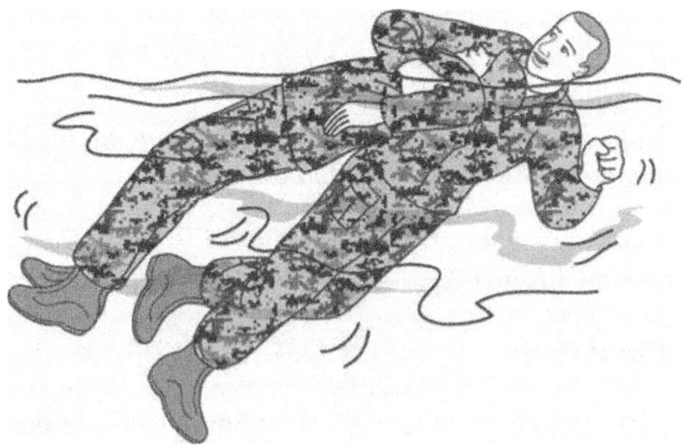

This procedure brings the victim's face and shoulders clear of the water, and typically the victim stops struggling. Sometimes, however, the victim will struggle during the swim to safety. If this happens, either tighten your grip on the victim or defend yourself with one of the techniques discussed on pages 2-25 through 2-30.

Tired Swimmer's Assist

You must maintain a 2 to 6 yards margin of safety from the victim at all times while you are getting into position to perform the tired swimmer's assist. Once behind the victim, you extend one arm, hand straight. Place your straight hand underneath the victim's opposite armpit. Maintain a 45 degree angle away from the victim, your arm is locked out. Assist the victim in propulsion until both of you reach safety.

Defense Against a Drowning Victim

DO NOT sacrifice your life in an attempt to save the victim. A struggling drowning victim poses great danger to anyone nearby. Driven by panic, the victim can grab you with great strength in an effort to climb out of the water. This can result in death for both you and the victim. The following techniques can be used to defend against a drowning victim, but the best defense against attack by the victim is to stay out of reach.

Block

The block prevents the victim from grabbing you if you have approached from the front. If the victim lunges toward you, react as follows:

> *Place one or both open hands against the victim's upper chest; being careful to avoid the victim's face, neck, and abdomen.*

Lean backwards and submerge rapidly. Keep your blocking arm(s) extended.

Swim underwater and away from the victim, and return quickly to the surface.

Stop 2 to 6 yards from the victim to reassess the situation.

Determine an appropriate course of action.

Wrist-Grip Escape/Wrist-Grip Escape Alternative

The wrist-grip escape is used when a victim grabs your arm or wrist. Submerge the victim quickly by reaching across with your free hand, pushing down on the victim's shoulder to submerge him, and kicking to propel yourself upward. While keeping the victim submerged with your hand on his shoulder, give three hard jerking pulls with your trapped hand in an attempt to break free from his grasp. Once free, swim clear of the victim and reassess the victim's condition.

If the wrist-grip escape is unsuccessful in freeing your hand, use the wrist-grip escape alternative: use your free hand to grab your trapped fist, rotate thumbs up, apply bone-on-bone contact with the victim's arm, pry your hand out of his grasp, and quickly swim away from the victim.

Front Head-Hold Escape

The front head-hold escape allows you to escape when you are facing a victim who is gripping you around your head and neck. To execute the front-head hold escape—

Take a quick breath and tuck your chin into your shoulder to protect your throat.

Clap your hands above your head (three times) to submerge instantly. This drags the victim below the water, lifts his arms from around your neck, and, typically, he releases his grasp in order to get back to the surface. If he doesn't release his grasp, apply pressure to the victim's brachial pressure points (which are located inside of the upper arm, above the elbow).

Thrust the victim's arms up and away.

Keep your chin tucked to protect your throat, and swim underwater away from the victim and return quickly to the surface at the ready position.

Stop 2 to 6 yards from the victim to reassess the situation.

Determine an appropriate course of action.

Rear Head-Hold Escape

The rear head-hold escape allows you to escape when a victim is gripping your head and neck from the rear. To execute the rear head-hold escape—

Take a quick breath. and tuck your chin down, turn your head to either side, and raise your shoulders to protect your throat.

Take a strong stroke, clap your hands above your head (three times), and submerge instantly. This drags the victim below the water and, typically, he releases his grasp in order to get back to the surface. If he doesn't release his grasp, apply pressure to the victim's brachial pressure points (which are located inside of the upper arm, above the elbow).

Thrust the victim's arms up and away.

Twist your head and shoulders until free.

Swim underwater away from the victim and return quickly to the surface.

Stop 2 to 6 yards from the victim to assess the situation.

Determine an appropriate course of action.

Administering First Aid/Rescue Breathing

To administer first aid/rescue breathing, remove the unconscious victim from the water when possible and stop any bleeding.

First, open the airway and check for breathing. If the victim is not breathing, give two breaths of air. Check for a pulse. If there is a pulse but the victim is not breathing, continue rescue breathing. If there is no pulse, start cardiopulmonary resuscitation (CPR). For detailed information on administering CPR and rescue breathing, refer to American Red Cross CPR skill cards.

WARNING

If the victim has no pulse and is not breathing, administer CPR immediately. If the victim does have a pulse but is not breathing, give rescue breathing only. If the victim has a pulse and is breathing, DO NOT give CPR—CPR could prove fatal.

Next, protect any wounds from exposure, and treat for shock.

Chapter 3
Treatment of Casualties and Avoidance of Dangerous Marine Life

To survive in the water, you will face many challenges other than just being able to swim. For example, injuries and/or fatigue can lead to drowning and exposure to the elements can lead to hypothermia or heat injuries and may require medical attention. Finally, your presence in the water may be seen as a threat to or food source for a variety of marine life and you may be attacked. This chapter advises you on how to minimize these risks in order to stay alive.

Drowning

Drowning is suffocation by a liquid. A drowning victim inhales water into the lungs or the throat closes by reflex so that little or no water can enter the windpipe. In either case, a victim can no longer breathe.

Symptoms

One symptom of drowning is that the victim may call for help and has an expression of dread or panic. But typically a victim that is active and drowning may not call for help because he is trying to conserve his air and will not speak.

Another symptom of drowning is when the victim thrashes at the water's surface. If the victim stops or grows calmer, he has likely been overcome by fatigue, hypothermia, or a lack of air. At this stage, the victim usually has 20 to 60 seconds before going under the water's surface.

Treatment

If the victim is not breathing, begin rescue breathing. Place the victim on his back, tilt head back to open airway, pinch the nose, and give two full breaths. If the victim does not inhale during the first two breaths, reposition his head and attempt two more breaths. Check for a pulse. If a pulse is present, but the victim is still not breathing, continue rescue breathing. If a pulse is not present, begin CPR. See MCRP 3-02G, *First Aid*, for rescue breathing and CPR details.

WARNING

If the victim has no pulse and is not breathing, administer CPR immediately. If the victim does have a pulse but is not breathing, give rescue breathing only. If the victim has a pulse and is breathing, DO NOT give CPR—CPR could prove fatal. Continue first aid until medical help arrives.

A victim who is not breathing and has no pulse may appear dead. DO NOT decide that death has occurred. Continue with the prescribed treatment. A corpsman or medical officer should decide whether the victim can or cannot be revived.

Hypothermia

Hypothermia is the abnormal lowering of the body's internal (or core) temperature to 95 degrees or below. It occurs when the body loses heat faster than it produces heat. A hypothermia victim loses the ability to move quickly, becomes mentally sluggish, slips into semiconsciousness, lapses into a coma, and dies when internal body temperatures drop too low.

The chilling effects of cold air, wind, or water can produce hypothermia. Water poses the greatest threat because it transfers heat 25 times faster than air. Depending on the water's temperature, a victim can succumb to hypothermia within a few minutes. The body's sudden contact with cold water can also set off a body reaction known as the mammalian diving reflex. This reflex can greatly increase survival time (especially for women and children) in or under cold water. The mammalian diving reflex shuts off blood circulation, except for the flow between the heart, lungs, and brain. The small amount of oxygen left in the blood and lungs is saved for the body's vital organs. This reflex has allowed people to survive being under cold water for an extended period of time. Therefore, a cold water drowning victim should be treated as if still alive even though the victim is not breathing, has no pulse, and may appear dead. DO NOT decide that death has occurred. Continue with the prescribed treatment. Victims of hypothermia can appear to be dead when they are not. A corpsman or medical officer should decide whether the victim can or cannot be revived.

WARNING

If the victim has no pulse and is not breathing, administer CPR immediately. If the victim does have a pulse but is not breathing, give rescue breathing only. If the victim has a pulse and is breathing, DO NOT give CPR—CPR could prove fatal. Continue first aid until medical help arrives. Check for a pulse for at least 45 seconds.

Symptoms

Once the body's core temperature drops, the victim will show one or more of the following symptoms:

- Violent and uncontrollable shivering as the body tries to warm itself.
- Slow or slurred speech.
- Disorientation or poor coordination.
- Loss of skin color.
- Blue and pinched lips.
- A slowing or stopping of shivering that progresses into a rigid torso and limbs.

Survival Time

A hypothermia victim's survival depends on the water's temperature and the time spent in the water. A small body build cools faster than a large build. Children cool faster than adults.

To increase your chance of survival in the water, utilize the HELP position described on page 1-37. Extra clothing and inactivity (remaining motionless in the water) can also increase your survival time.

Treatment

A hypothermia victim must be warmed to prevent further heat loss; therefore, treatment should begin as soon as possible. Consciousness of the victim determines the treatment that should be pursued. See MCRP 3-02G for specific treatment information. The following treatment procedures are recommended:

- If the victim is conscious, give the victim warm fluids. Give candy or sweetened foods to a victim who is able to eat.
- If the victim is unconscious, place him on his back with his head tilted back to ensure an open airway.
- DO NOT massage the victim. Massage can break blood vessels and create swelling, internal pressure, and blocked blood circulation.
- DO NOT give alcohol to the victim. Alcohol lowers the victim's body temperature.
- Shock is a possibility, treat accordingly.
- Seek medical help immediately.

If you are able to remove the victim from the water, apply the following steps when possible:

- Get the victim into shelter.
- Remove the victim's wet clothing.
- Put the victim in dry clothing.
- Place the victim in a sleeping bag if one is available. It may be necessary to place another Marine in the sleeping bag with the victim.
- Place as much insulation as possible between the victim and the ground.
- Use hot water bottles, electric blankets, or blankets heated in an oven or by a campfire to warm the victim's neck, groin, and the sides of the chest.

CAUTION: DO NOT apply heat to extremities.

Heat-Related Injuries

Heat-related injuries include heat cramps, heat exhaustion, and/or heat stroke. They can occur when a Marine—

- Is exposed to extreme heat, such as from the sun or a combination of high air temperatures and water temperatures.
- Does not wear proper clothing or gets overexposed to the sun.
- Becomes dehydrated.

With heat cramps, muscles may cramp in the arms, legs, and/or stomach. The victim may sweat excessively. To treat, create improvised shade for the victim and have him drink water.

With heat exhaustion, the victim sweats heavily; presents pale, moist, cool skin; and complains of a headache, weakness, dizziness, and/or loss of appetite. A victim may also experience heat cramps, nausea, vomiting, an urge to defecate, chills, rapid breathing, confusion, and a tingling sensation in the hands or feet. To treat, pour water over the victim and fan him to speed up the coolant effect of evaporation, have the victim drink freshwater, and attempt to provide shade.

With heat stroke, the victim stops sweating, which results in red, flushed, hot, dry skin. The victim may first experience headaches, dizziness, nausea, fast pulse and respiration, and/or seizures and mental confusion. The victim can become unconscious and die if not treated quickly. To treat, create improvised shade for the victim. Have him drink water and elevate both legs if possible.

Burns

Apply a field dressing (or the cleanest material available) to the burn. Give sips of water to a casualty who is conscious and not nauseated. When treating burn casualties—

- DO NOT remove clothing stuck to the burns.
- DO NOT break any blisters.
- DO NOT apply grease or ointments to the burns.

For electrical burns, check for both an entry and exit burn from the passage of electricity through the body. An exit burn may appear on any area of the body, not necessarily opposite the entry burn.

For burns caused by wet or dry chemicals, flush the burns with large amounts of water and cover with a dry dressing.

For burns caused by white phosphorus, flush the area with water, then cover with a wet material, dressing, or mud to exclude air and keep the white phosphorus particles from burning.

For laser burns, apply a field dressing.

Common Medical Problems Associated with Sea Survival

Seasickness

Seasickness is the nausea and vomiting caused by the bobbing motion created by the wave action of a flotation device. Seasickness can result in—

- Dehydration and exhaustion.
- A loss of the will to survive.
- Others becoming seasick.
- Unclean conditions.

To treat seasickness—

- Wash both the Marine and the flotation device to remove the sight and odor of vomit.
- Keep the Marine from eating food until the nausea is gone.
- Have the Marine lie down and rest.
- Give the Marine seasickness pills if available. If the Marine is unable to take them orally, the pills should be inserted rectally for absorption by the body.

Saltwater Sores

Saltwater sores occur when skin that has abrasions or is cut is exposed to saltwater. The sores may form scabs and pus. Do not open or drain. Flush the sores with freshwater, if available, and allow to dry. Apply antiseptic, if available.

Blindness/Headache

Irritants or the effects of the sun's rays reflecting off the water can cause temporary blindness or headaches. If flames, smoke, or other irritants get in your eyes, flush the eyes immediately with saltwater, then with freshwater, if available. Apply an ointment, if available. Bandage both eyes for 18 to 24 hours or longer if the damage is severe. If glare from the sky and water causes your eyes to become bloodshot and inflamed, bandage the eyes lightly. Try to prevent this problem by wearing sunglasses or goggles with a sunglass insert.

Constipation

This condition is a common problem associated with dehydration. For constipation, do not take a laxative if it is available; this causes further dehydration. Drink freshwater, if available.

Sunburn and Dehydration

The sun's rays reflect at all angles off the waves of the water; therefore, sunburn and dehydration are serious problems in sea survival. Try to prevent sunburn by—

- Erecting an improvised canopy, with available floating materials, to provide shade.
- Wearing your soft cover or using a cloth, such as a handkerchief, to cover your head.
- Covering your skin with sunscreen or lip balm from your first aid kit. Your lips, nostrils, eyelids, the backs of your ears, and the skin under your chin sunburns easily. If enough sunscreen cream is available, all exposed skin should be covered.

Dehydration is caused by the loss of the body's vital fluids. Dehydration in saltwater may result from a combination of factors such as a lack of water, the effects of saltwater on skin tissue, sunburn or vomiting from seasickness, and other causes. Sleep and rest and reduced water and food intake are the best ways of enduring periods of exposure. The following measures will delay the effects of dehydration:

- DO NOT drink saltwater.
- DO NOT drink urine.
- DO NOT drink alcohol.
- DO NOT smoke.
- DO NOT EAT unless water is available.

Dangerous Marine Life

You may see many types of marine life around you, and some are more dangerous than others. Generally, sharks are your greatest danger, followed by barracudas. However, most marine life will not deliberately attack a human. The most common injuries from marine life are wounds from bites, stings, or punctures. With the exception of sharks and barracudas, most injuries are a result of either trying to catch game or from contact abrasion with marine life. To treat an injury that results from a dangerous marine animal bite—

- Pack the wound with gauze and then apply a pressure bandage, if available.
- Treat for shock.
- Obtain medical attention as soon as possible.

For a more detailed information on the treatment of casualties and dangerous marine life, see MCRP 3-02F and MCRP 3-02G.

Sharks and Barracudas

Only about 20 percent of all shark species are known to attack people. Sharks have an acute sense of smell, and the smell of blood in the water will draw them to their prey. They are also very sensitive to any abnormal vibrations in the water; therefore, the sound caused by a struggling swimmer or underwater explosions will attract them.

Barracudas are bold and inquisitive fish. They have been known to attack men who are wearing shiny objects. Barracudas may charge at lights or shiny objects at night.

A group of swimmers can maintain a 360 degree watch while in the water. Therefore, to protect yourself from sharks and barracudas, stay with other swimmers. The group members can either frighten away or fight off sharks or barracudas better than one person can. Keep all your clothing on, including footwear. Historically, sharks attack unclothed individuals in groups first, mainly in the feet. Clothing also protects you against abrasions from a shark's tough skin should the shark brush up against you. Avoid urinating heavily, let urine dissipate between discharges. If you must defecate, do so in small amounts and toss it as far away from the group as possible. Do the same if you must vomit.

If attacked, the use of firearms by swim sentries should be used with extreme caution because of the risk of injury to other swimmers. If unarmed or unable to make an improvised weapon, kick and strike the shark. Avoid using your bare hands to strike the shark, injury can result to your hands due to a shark's tough skin. Target areas on a shark are the gills, eyes, and underbelly. Blows to the snout are also not recommended because a shark will tilt its head up and thrust its jaws forward when biting.

Sea Snakes

Sea snakes are venomous and sometimes found in mid-ocean. They are unlikely to bite unless provoked. Your best protective measure is to avoid them.

Poisonous Fish

Many reef fish are toxic, and can kill you if eaten. Generally, poisons are present in all parts of the fish, but especially concentrated in the liver, intestines, and eggs.

Turtles and Moray Eels

Turtles and moray eels normally inflict minor bite wounds. Treat this type of injury by cleaning the wound.

Corals

Coral, dead or alive, can inflict painful cuts. Clean all coral cuts thoroughly. DO NOT use iodine to disinfect any coral cuts. Some coral polyps feed on iodine and may grow inside the flesh.

Jellyfish, Portuguese Man-of-War, Anemones, and Others

This group of marine animals inflicts injury by stinging their victims with their tentacles. Contact with their tentacles produces burning pain, a rash, and small hemorrhages on the skin. Shock, muscular cramping, nausea, vomiting, and respiratory distress may also occur. Gently remove the clinging tentacles with a towel and wash or treat the area. Use diluted ammonia or alcohol and talcum powder to treat the injury if available.

Spiny Fish, Urchins, Stingrays, and Cone Shells

These animals inject their venom by puncturing the skin with their spines. General signs and symptoms include swelling, nausea, vomiting, generalized cramps, diarrhea, muscular paralysis, and shock. Deaths are rare. Treatment consists of soaking wounds in hot water, if available, to deactivate heat-sensitive toxins.

Chapter 4
Negotiating Water Obstacles

Marines face water obstacles in saltwater, freshwater, and brackish water (where freshwater and saltwater meet). These water environments differ considerably and pose distinct problems for Marine tactical units and swimmers. Saltwater obstacles include tides, surfs, and currents. Brackish water obstacles include back bays. Freshwater obstacles include rivers and canals.

Tides

Tides are periodic changes in the surface levels of oceans, bays, gulfs, inlets, and rivers. The Moon's and Sun's gravitational pulls cause tides. Depending on the situation, tides can either help or hinder Marines in their endeavors to conduct amphibious or riverine operations. Tides can provide sufficient depths to allow for the passage of landing craft or boats over reefs, trees, and other underwater obstructions. On the other hand, tides can render a river fording site unusable to tactical vehicles. Direction, level of change, and amount of change determine tidal nomenclature.

Tides that show change in direction are flood tides and ebb tides. Rising tides are known as flood tides. Falling tides are known as ebb tides.

Tides that show extreme levels of change are high tides and low tides. High tide is the period when water is at its greatest depth. Low tide is the period when water is at its most shallow depth.

Tides that show the least and most amount of change are neap tides and spring tides, respectively. Neap tides have the least amount of change in water levels between high and low tide. Neap tides occur at the half moon, when the Sun and the Moon are aligned at a 90-degree angle with the Earth. In this position, the Sun's and Moon's gravitational pulls offset each other. Spring tides have the highest floods and lowest ebbs. Spring tides occur at or shortly after the new moon or full moon when the Sun, Moon, and Earth are approximately in line. In this position, the Sun's and Moon's gravitational pulls are combined.

Surf

Waves break upon entering shallow water and create surf. The offshore area where waves break is the surf zone, which can present many hazards. Breaking waves often trap air bubbles and create a foamy appearance. Bubbles lower the water's density and decrease buoyancy. Move through foamy surf as quickly as possible. The type of wave determines your survival technique.

Wave action moves you toward shore. Lie on your back or side with your head pointing in the direction of the beach and your feet pointing into the waves.

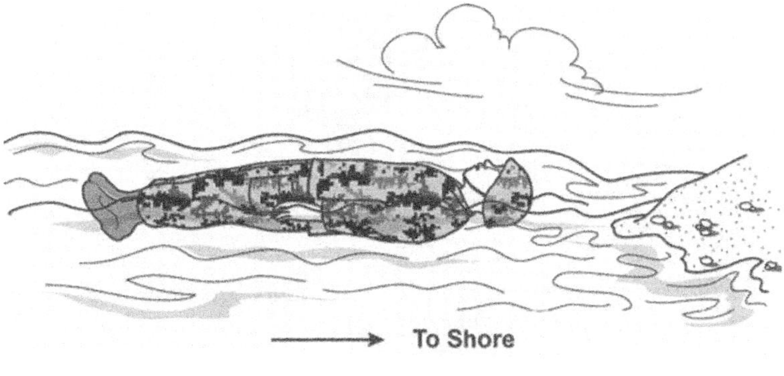

As one wave approaches the beach, another drains away; relax and do not swim against the draining water. When a new wave is within about 3 yards, start swimming toward shore. Continue to swim until the wave lifts you and moves you toward the beach. Once the wave loses forward momentum, relax and repeat the cycle. If nearing rocks, turn your body and approach feet first to reduce the chance of striking your head and arms.

Plunging Waves

A plunging wave is a breaker whose crest curls forward and falls ahead of its base. Because of its power and underwater turbulence, a plunging wave poses the greatest surf threat. If caught in a plunging wave, you can be pulled underwater and pitched about violently. This can cause you to panic, which can increase your chance of drowning. Perform the following steps to escape a plunging wave:

Tuck into a ball with your head against your knees and your forearms locked around your legs, just below the knees.

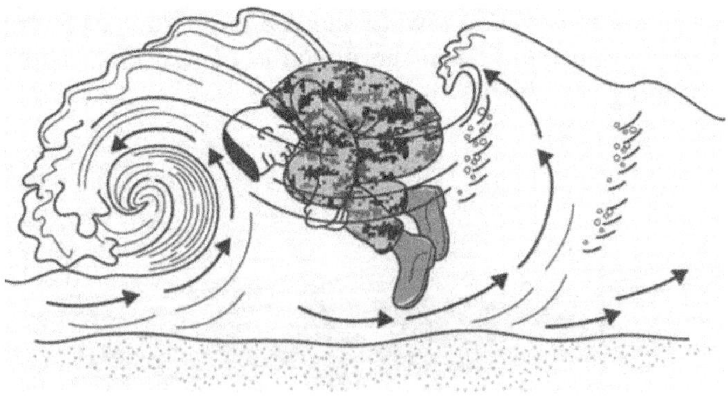

Relax in this position until the turbulence subsides and you float to the surface. This can take 30 seconds or more.

Swim toward shore.

NOTE: If threatened by another plunging wave, dive underwater into the wave.

Spilling Waves

A spilling wave does not break. Instead, its crest slides forward without curling. A spilling wave creates less turbulence and poses less of a threat than a plunging wave. If caught in a spilling wave, relax and let the wave carry you to shore.

Surging Waves

A surging wave occurs on a beach with a steep underwater gradient. It never really breaks, but the crest rises while the base slides up the beach with great force and speed. Once the wave reaches its highest point on the beach, it rushes back as quickly as it surged forward. If you are standing on the bottom when a surging wave advances or retreats, the wave can knock you off your feet and pull you into the surf zone. Do not try to stand or walk on the bottom. Swim toward the beach as soon as possible.

Currents

Offshore Currents

An offshore current occurs outside the surf zone. Typically, it occurs at bay entrances, in island channels, and between islands and the mainland. An offshore current flows parallel to or away from shore. If the offshore current is created by tides, its current strength and direction vary at different times of the day.

If caught in an offshore current, you may be carried in a direction you do not want to go. DO NOT try to swim directly to safety. If the current is moving directly away from the shore, relax and wait until the current dies out or turns toward land. Once the current subsides, use any survival stroke to swim toward shore. If the current is moving parallel to shore, use any stroke to move at an angle across the current and toward shore.

Rip Currents

A rip current occurs when waves pile water against the shore faster than the water can drain. The water flows rapidly along the beach until it is deflected seaward by a bottom obstruction. Then the rip current flows through the surf zone and into open water at a speed of up to 2 knots. This action can cut deep trenches in the sand. A rip current dies out once it reaches open water (usually within a few hundred yards of the shore).

A rip current can pull you out to the open sea. If caught in a rip current, DO NOT try to swim against the current. A rip current moves faster than most people swim, and it is impossible to swim to shore once caught in it. Relax and stay afloat until the current runs out. Once the current subsides, use a survival stroke to move parallel to the shore until you are out of the current, then begin swimming toward shore.

Littoral Currents

A littoral current occurs when a wave breaks against a beach at an angle. This current flows parallel to the shoreline and does not pose a great threat. If caught in a littoral current, use the combat travel stroke to swim across it at an oblique angle.

Back Bays

Once on the beach, you often face one or more rows of low hills called dunes. Behind the dunes, you may encounter a low-lying stretch of ground thickly covered with scrub trees and bushes. This area gives way to wetlands known as back bays. Back bays consist of muddy islands that are almost submerged during flood tides and separated by channels of brackish water of varying depths. Channel bottoms usually contain soft mud. Back bays pose major obstacles to vehicular traffic.

Infantry can cross back bays, but only with great effort. If crossing back bays by foot, consult detailed navigation charts and use the following guidelines to plan your route:

- Avoid water less than waist deep; walking in shallow water or soft mud is extremely tiring.
- Avoid back bay islands; these low-lying islands are usually too muddy to support foot traffic.
- Seek out deep water; floating with a pack is less tiring than walking through shallow water or soft mud.
- Seek out sand, shell, gravel, or stone bottoms; these firmer bottoms generally ease travel and help conserve energy.

Rivers and Canals

A river is a large, natural stream of water that empties into a larger body of water. The slope of the riverbed and the volume of water in the river determine its current.

Canals resemble small rivers or streams in their width and depth, but usually lack any significant current. Climbing out of these waterways can be difficult if the canal is flanked by steep banks.

Chapter 5
Fording Waterways

WARNING

Fords are dangerous. Cross as quickly as possible.

A ford is any site in a river, stream, or canal where the water is shallow enough for troops or vehicles to cross without using flotation devices. Canal bottoms are usually too soft to support fording vehicles, and wading infantry frequently stumble.

The tactical situation dictates the location of the fording site. Seek fords that are protected from enemy observation and that allow for adequate supporting fires. A night fording takes at least one and a half times as long as a daylight fording.

NOTE: In this chapter, the term "river" refers to rivers, streams, and canals.

Selection of a Ford Site

The following table identifies desirable fording characteristics:

Characteristics	Comments
Concealment	The ford hides personnel and vehicle movement from enemy observation.
Accessibility	The ford should have low banks with gentle gradients. This allows a free flow of traffic at both the entrance and the exit.

Characteristics (Continued)	Comments (Continued)
Slow Current	The ford's current should not exceed 1.5 meters per second if possible.
Firm Footing	The ford's bottom, entry, and exit composition should be firm enough to support traffic. Do not drive a vehicle over any bottom composition that a 2-inch diameter stick can be pressed into more than 1 or 2 inches.
Gently Sloped Channels	The ford's entry and exit points should be gently sloped. If possible, locate a portion of the stream where the channel is not actively shifting.
Depth	The fording depth is less than or equal to the least capable vehicle.

Determine the Slope

Units move into and out of water faster and more quietly if entry and exit points are not steep or muddy. Slope is the amount of change in ground horizontal distance (run) and in vertical elevation (rise) from one point to another. Slope is usually expressed as a percentage. You can use a clinometer, map, or line of sight and pace to measure the percentage of a slope.

Clinometer

A clinometer measures percentage of a slope. It is a component of the M9 armored combat excavator, and organic to many Marine engineer units. However, engineers are also taught (and often use) field-expedient methods to determine slopes and gradients.

Map

A map measures the horizontal distance along a desired path. Determine the difference in elevation between the path's starting and ending points. From a map's scale, you can determine the distance between two points. From a map's elevation lines, you can determine the difference in height between the same two points. Both figures must be in the same unit of measure (e.g., feet, meter). Divide the elevation (rise) by the distance (run) and multiply by 100.

Rise = 165 - 120 = 45 meters

Run = 200 meters

% Slope = $\frac{45}{200}$ **x 100 - 22.5%**

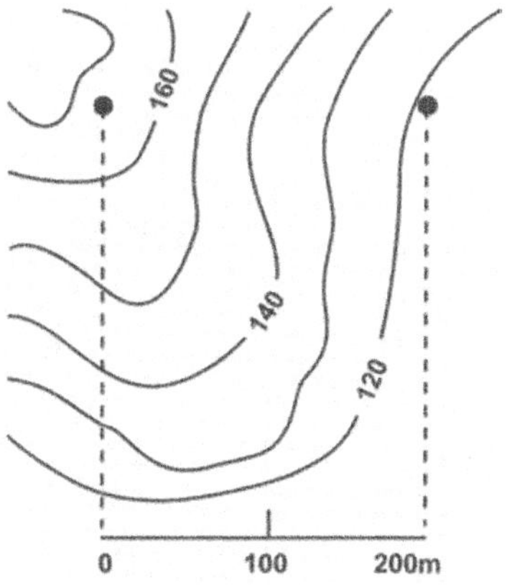

Line of Sight and Pace

To determine line of site and pace, stand at the bottom of the slope, keep your eyes level, pick a spot on the slope, then pace the distance. The number of paces multiplied by a standard measure of 0.75 meter determines the run. The eye-level height (usually 1.5 to 1.7 meters) determines the rise. Repeat this procedure until you have covered the entire distance you want to measure for each spot (vertical and horizontal). Add the vertical distances to provide total rise and the horizontal distances to provide total run.

Determine the Current Speed

Current speed increases as channels narrow. It may be necessary to locate a wider ford location to obtain a slower stream current. The following steps are used to calculate the speed of the current:

Determine points A and B along a channel. Then measure the distance between those two points.

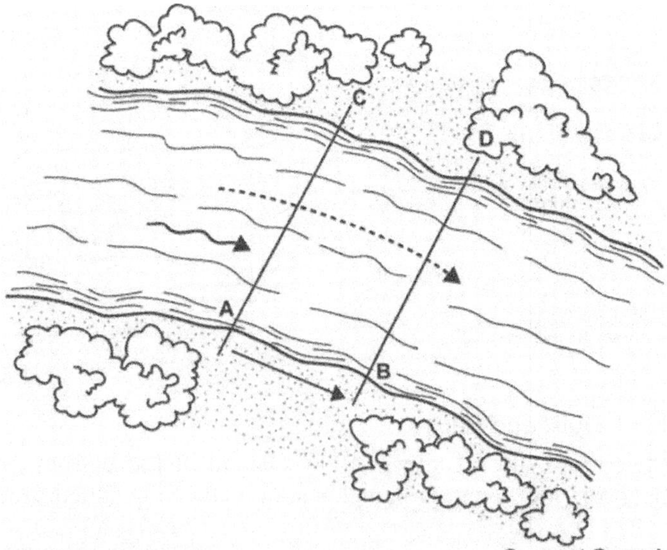

$$\frac{\text{Distance (meters) Between Points A and B}}{\text{Time (seconds) Between Points C and D}} = \text{Current Speed (meters per second)}$$

Sight directly across the water from points A and B to locate points C and D.

Throw a floating object (e.g., a stick) upstream from points A and C. Observe the object as it floats toward points B and D.

Subtract to find the time it takes for the object to float from start to finish.

Do not attempt to swim across currents that are moving faster than 1.5 meters per second. Equivalents of this speed include—

- Quick-time march rate of 120 counts per minute with one, 30-inch step at each count.
- 5 feet per second.
- 3.5 miles per hour.
- 5.5 kilometers per hour.

Measure River Width

A river's width can be estimated from the width of its symbol on a scaled topographic map. If this is not possible, use the following compass techniques:

Stand at the water line (A).

Shoot an azimuth to a point on the opposite bank (B).

Move upstream or downstream until you are at a point (C) where you can shoot an azimuth 45 degrees larger or smaller than the original azimuth.

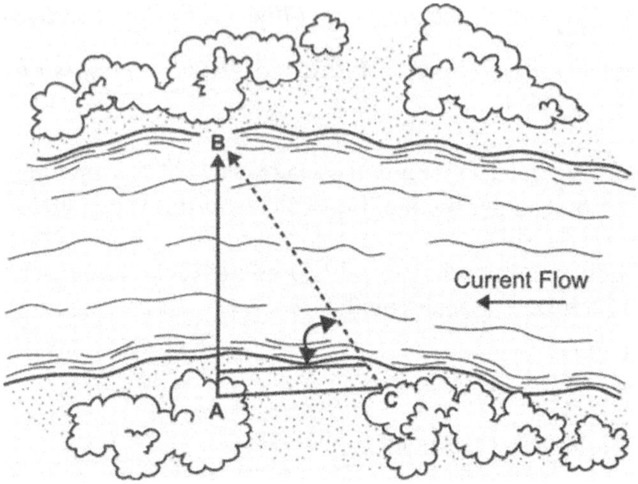

Measure the distance between points A and C. The distance calculated equals the river's width.

Calculate Downstream Drift

A river's current causes personnel and equipment to drift downstream. If personnel and equipment are aimed straight across the river, they will sideslip downstream as they move across the current to the other shore. Therefore, personnel and equipment crossings must compensate for the effects of a river's current and entries are usually made upstream of the desired exit point. Use the formula below to calculate downstream drift.

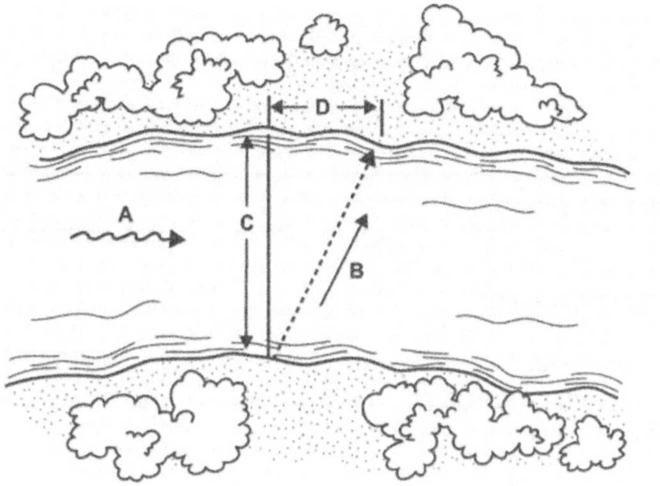

$$\frac{\text{Current Speed (A)}}{\text{Crossing Speed (B)}} \times \text{River Width (C)} = \text{Downstream Drift (D)}$$

NOTE: The crossing speed for a swimmer across a river may vary but is generally limited to 1 meter per second. All measurements must be in the same unit of measure (e.g., meters, feet).

The Buddy System

Whenever a Marine unit must enter into or operate on the water, a "buddy system" is employed in which every Marine is paired with a swimming partner. The buddy system matches an experienced swimmer with a weak swimmer. The experienced swimmer assists and encourages the weaker swimmer and bolsters confidence during night crossings. If a unit has an odd number of Marines, place the extra person with another pair to form a three-person team.

Water Crossings

Care of Weapons

Marine infantry weapons and munitions are designed to be able to operate after immersion. However, protect your weapons from moisture whenever possible.

A gas-operated weapon can malfunction if water travels down the barrel and enters the gas tube. To protect the gas tube—

Close the weapon's bolt before entering the water.

Seal the muzzle with a condom, balloon, plastic spoon wrapper, or other form of waterproof material.

Tie or melt the protective cover to create a watertight seal.

When the muzzle's protective cover is no longer needed, remove it. Open the bolt and inspect the barrel. If the tactical situation permits, swab excess moisture from the barrel. Test fire automatic weapons, if possible. Field strip and clean weapons as soon as possible. If time does not allow for a complete inspection, rinse inaccessible areas with small amounts of diesel fuel, then dry.

ISOMAT Raft

Construction of an ISOMAT raft is time-consuming. This type of raft should not be employed as part of an attack, but used for logistical purposes (e.g., evacuating stretcher cases, transporting supplies). Use the following steps to build an ISOMAT raft:

Wrap ISOMAT sleeping pads around sturdy sticks.

Use parachute cord and square knots to tie the pads securely in place and to lash stick ends together in a rectangle.

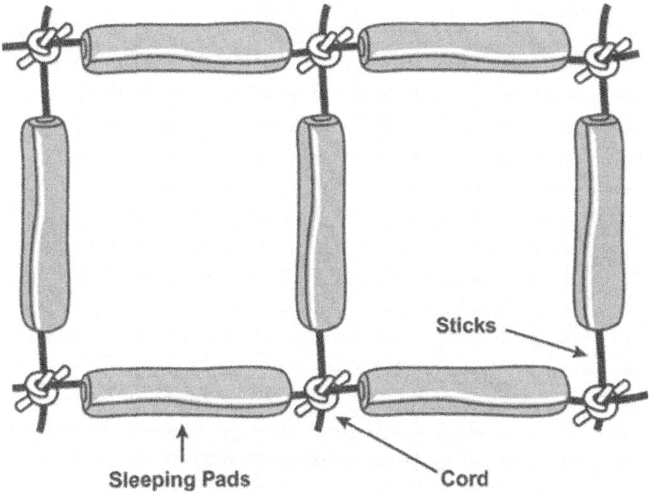

Note: The ISOMAT raft pictured can support several hundred pounds. However, the cargo will get wet if not properly waterproofed.

Poncho Raft

A poncho raft can support two Marines and their equipment and is well suited for long crossings. Use the following steps to build a poncho raft:

Inspect two ponchos and ensure they are serviceable.

Lay one poncho flat on the ground, with the hood-side up.

Cinch the hood tightly to form a gooseneck or tie in a knot.

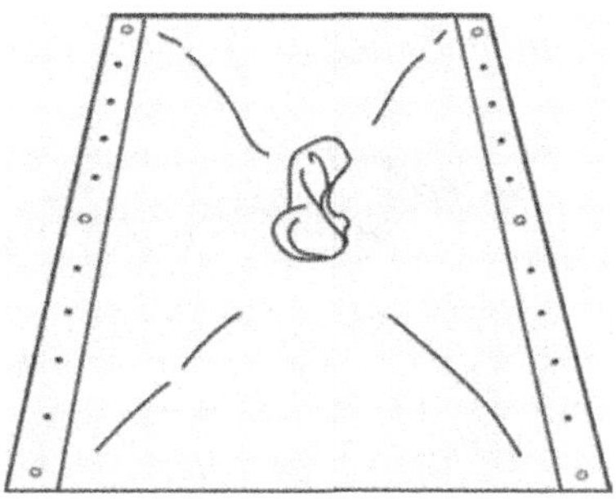

Pad sharp edges of equipment and place the equipment in the center of the poncho.

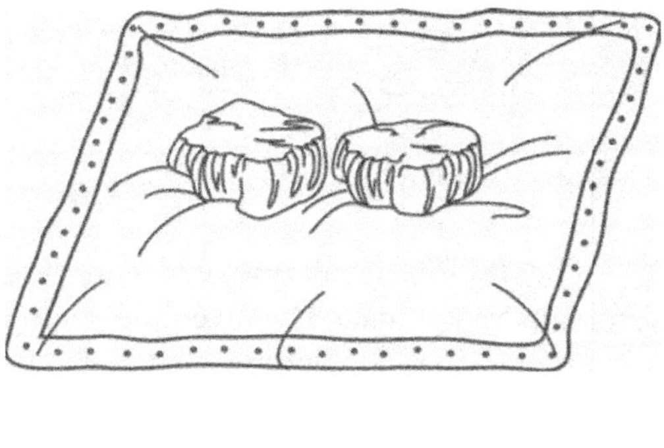

Place the second poncho over the equipment, rubber side up, and hood facing down.

Snap the edges of the two ponchos together.

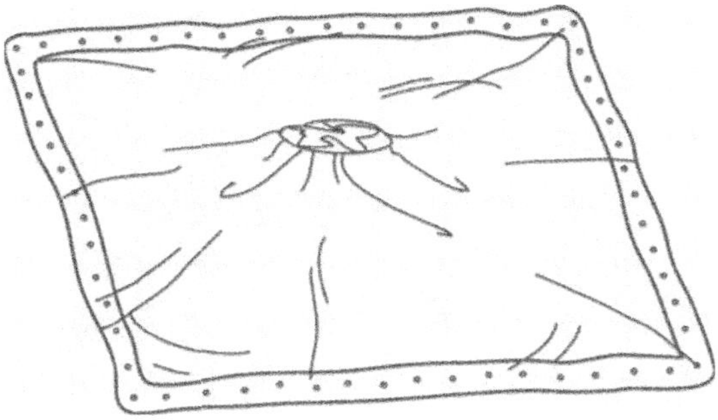

Roll the edges toward the equipment.

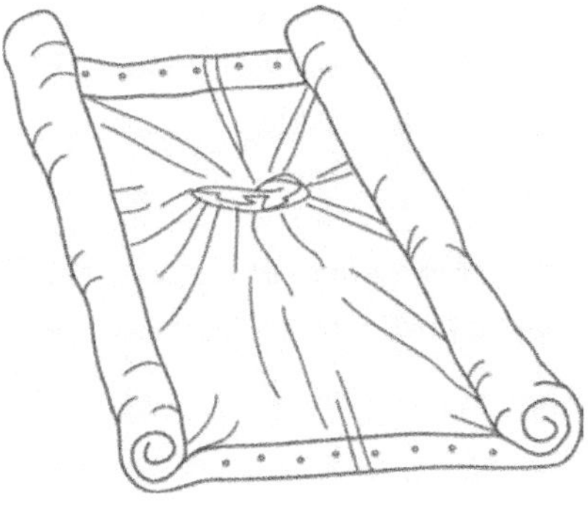

Roll the edges into pigtails and tie them off.

Pull the pigtails together over the top and lash them securely.

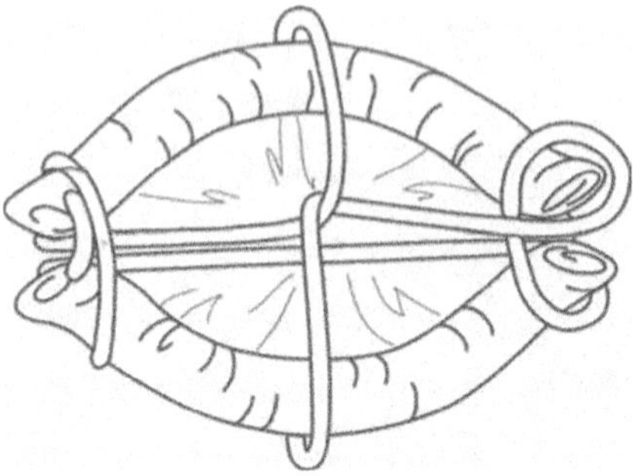

Protect the raft from brush punctures while placing it in the water. Swim across the water obstacle while security elements are covering the far shore.

Construction of a Pack Raft

You will need two waterproofed packs and two M16A2 service rifles. The following steps are required to construct a pack raft:

Place two packs side-by-side with the pack frames on the deck. The tops of the packs are opposite of each other.

Loosen the main compartment straps on both packs.

Insert one rifle on each end between the straps and the packs. The muzzles are opposite of each other. The rifles serve as one means to secure the packs together. Place the front sight post under the top flap.

Tighten the straps so that the rifles and packs are secure.

Take the excess strap on the inner side of each pack and secure it to the opposite pack to better secure the two packs together.

Take the excess straps on the outer sides of the packs and use those straps as safety lashing for the rifles.

Tuck the excess straps and check to make sure the rifles and packs are secure.

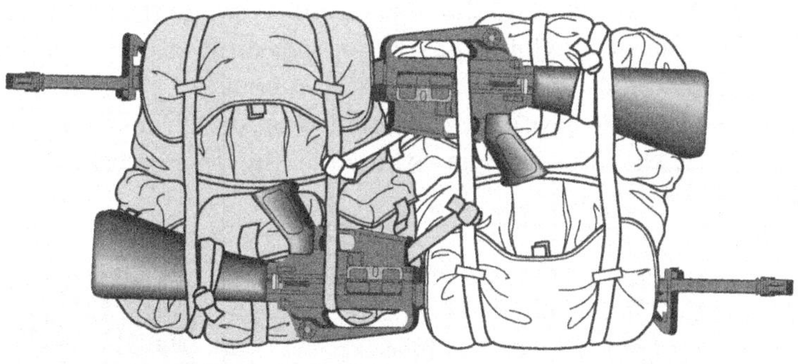

Single-Rope Bridge

A single-rope bridge offers a temporary and quick way to cross small rivers. It also provides extra security while crossing swift waters. At night, it prevents straggling, and guides units precisely from one side of the river to the other side. If crossing a river at night, plan for at least one single-rope bridge.

If your unit is crossing a river with swift currents or water depths above 4 feet, the unit is carrying sufficient rope to span the crossing site, and the tactical situation permits, secure the rope on near and far banks to provide a hand-hold for crossing Marines. This reduces the time required for the entire unit to cross and provides a degree of comfort/confidence for poor swimmers. Use a squad-sized bridge team to construct a single-rope bridge. Station several strong swimmers at the water's edge to help anyone who has trouble crossing.

Nylon rope is normally coiled in 120 foot lengths. It is 0.6 inches in diameter and has a breaking strength of about 3,840 pounds. Over time, a nylon rope can stretch to as much as one-third more than its original length and stretching weakens the rope. If the rope is stretched, discard the rope or use it for light tasks. To prolong the life of a nylon rope, do not step on it or drag it on the ground. Pad the rope in places where it contacts rocks or sharp corners. Do not leave the rope knotted or stretched longer than necessary. Dry rope as soon as possible. Single-rope bridge construction is as follows:

Tie a sling rope around your waist using a square knot and two, separate half hitches. See the appendix for detailed information on knots.

Attach a locking steel carabiner to the sling rope.

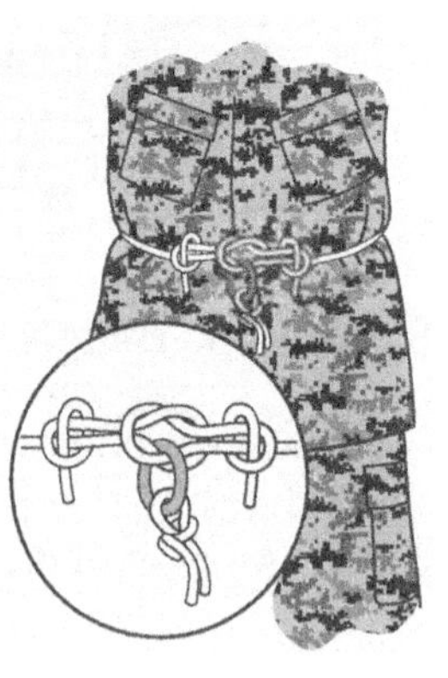

Tie a bowline knot in the running end of the bridge rope and attach it to the carabiner.

Temporarily secure the other end of the rope to a tree on the near shore.

Enter and cross the water.

NOTE: Carry only your weapon and ammunition.

Exit the water on the opposite shore.

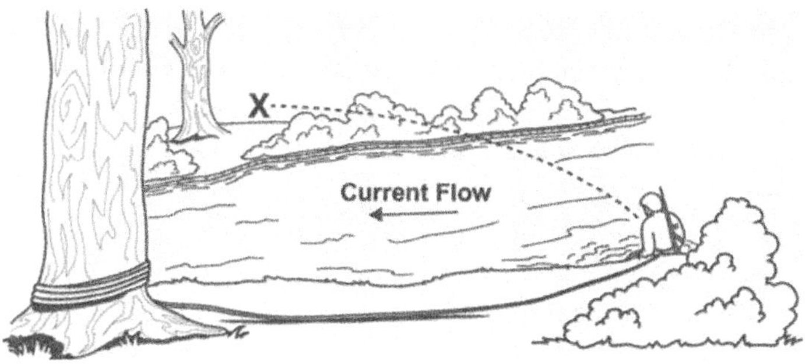

Prepare your weapon for use. Unhook the bridge rope from the carabiner at your waist, and tie the bridge rope to a sturdy tree using a round turn and two half hitches.

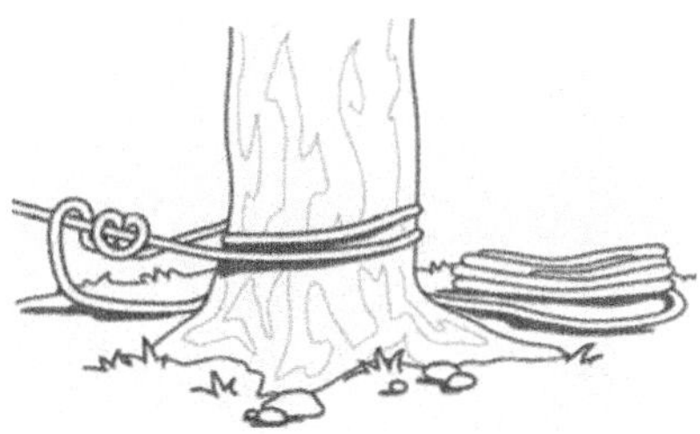

Conduct a box reconnaissance of the opposite shore.

On the near shore, have another Marine prepare to tighten the rope. That Marine should place a transport tightening system in the bridge rope by tying a double butterfly knot and placing two carabiners in the butterfly. See the appendix for detailed information on knots.

The Marine should pass the running end of the bridge rope around the downstream side of the near shore anchor point and through the two carabiners.

Pull the butterfly knot approximately one-third of the distance across the river.

Secure the bridge rope to an anchor point using a round turn and two half-hitches.

On the near shore, the Marine helping you should pull the slack out of the bridge rope until the butterfly knot is back on the near side. The bridge rope is then tied off against itself using two half hitches with a quick release in the last half hitch.

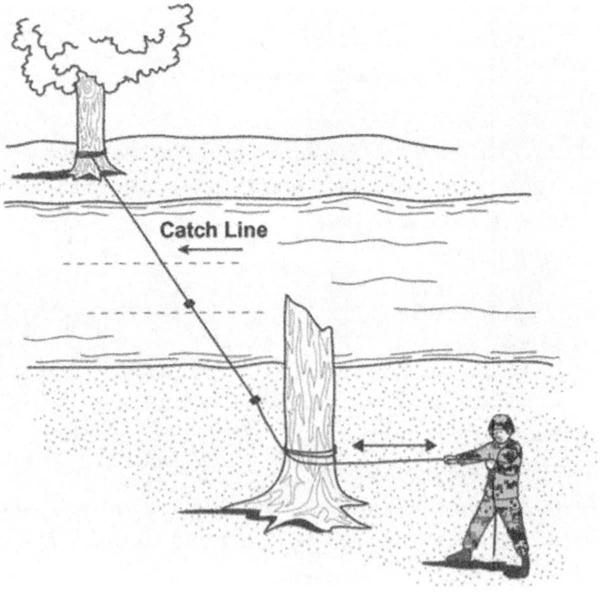

Note: The single-rope bridge must be as tight as possible so it will not sag when used.

If you lose your footing and fall into the water, swim with the current to the closest shore. Swimming against the current is dangerous and quickly causes fatigue.

High and Dry Crossings

If the single-rope bridge is high enough, suspend yourself below the single-rope bridge and above the water. Use the following steps to suspend yourself from a single-rope bridge and then pull yourself across the water:

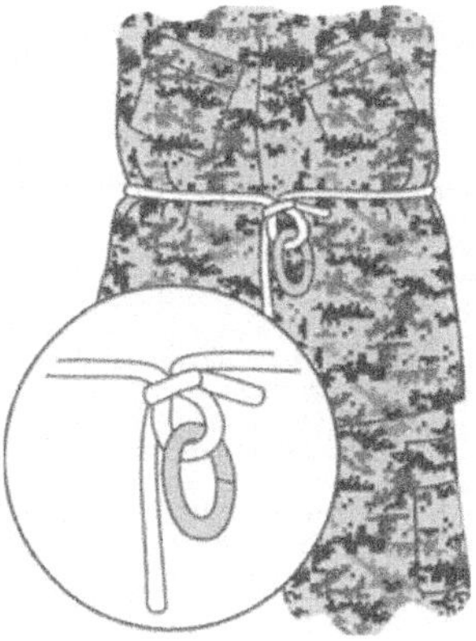

Tie a sling rope around your waist using a bowline. Ensure that the knot is tight.

Attach a carabiner through the bowline's loop. The carabiner's gate faces up.

Secure your helmet chin strap.

Face the single-rope bridge with your left shoulder toward the far shore.

Grasp the bridge rope in both hands.

Swing your body beneath the single-rope bridge with your head toward the far shore. Cross your ankles above the bridge rope.

Arch back until the carabiner contacts the bridge rope. Connect the carabiner to the bridge rope. Allow the carabiner to bear your body's weight.

Pull yourself across the single-rope bridge, hand over hand, to the far shore.

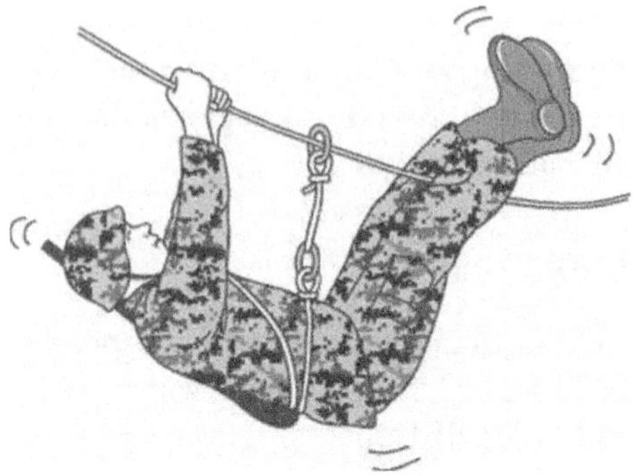

Swift Current Crossings

A single-rope bridge prevents being knocked down and swept away by a swift current. Use the following steps to move through a swift current:

Tie one end of a sling rope around your waist using a bowline.

Tie the running end of the sling rope in another bowline, and attach a carabiner to the bowline's loop.

Step up to the bridge. Face upstream.

Hook the carabiner to the single-rope bridge.

Walk sideways into the river while grasping the bridge rope in both hands.

Use the single-rope bridge for balance and remain standing, if possible.

Continue to move sideways through the river to the far shore.

Slow Current Crossings

If you face little or no current, it is not necessary to hook up to a bridge rope with a carabiner. Lie on your back in the water beneath the single-rope bridge. Support your body weight with your waterproof pack. Use the bridge rope and pull yourself hand over hand across the river.

Removal

If you are the last Marine waiting to cross, pull on the standing end of the rope to release the knot, then tie the rope around your waist using a bowline. The Marines on the far shore will pull you through the water.

Appendix
Knot Tying

Square Knot

The square knot is used to secure two ropes of equal diameter together so they form one continuous rope that will not slip.

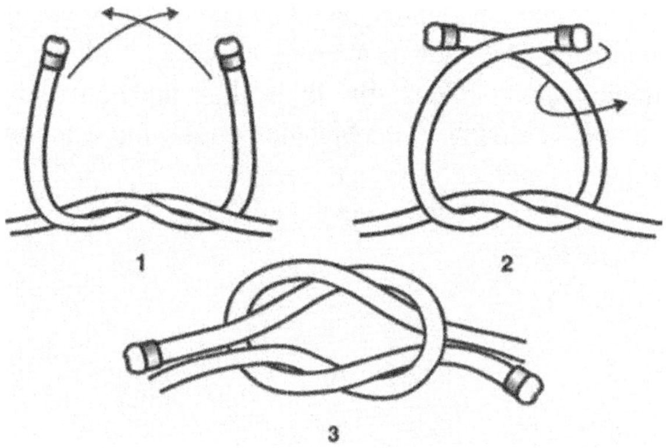

Bowline

The bowline forms a loop that will not tighten or slip under strain. It is easily untied.

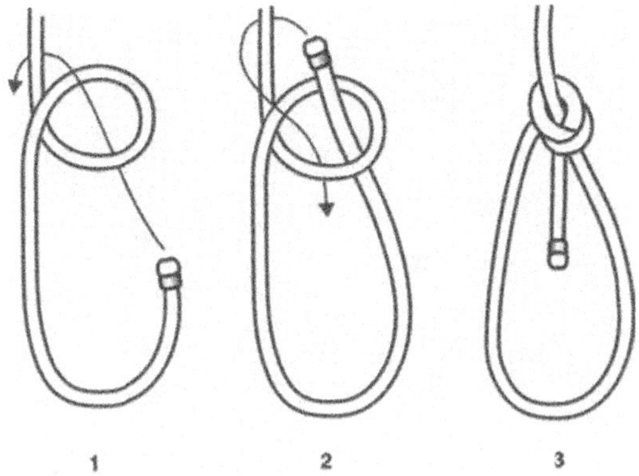

Hitches

Hitches are used to dress (prepare) knots and secure loose ends.

Half Hitch

The half hitch is used to tie a rope to a tree or to a larger rope. It will hold against a steady pull, but is not a secure hitch. It is frequently used to secure the free end of a rope.

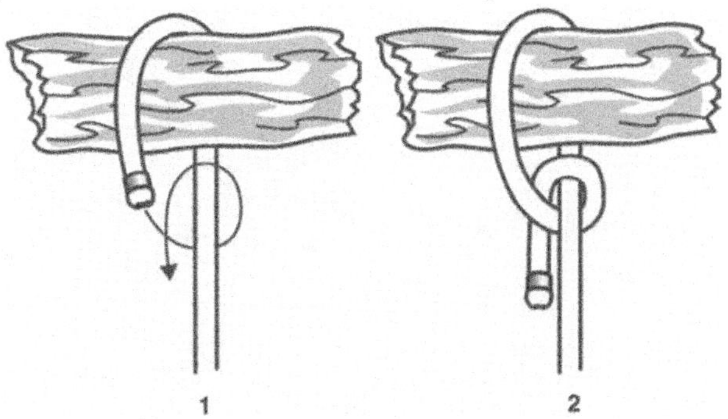

Two Half Hitches

Two half hitches can be used to secure the running end of a rope.

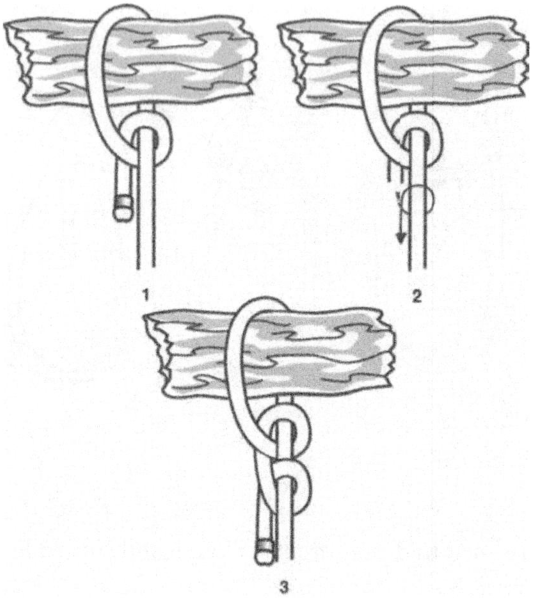

Round Turn and Two Half Hitches

A round turn and two half hitches can be used to fasten a rope to a tree. This hitch does not jam.

Butterfly Knot

The butterfly knot is used for anchor lines. Tied properly, it will not tighten on itself to the point that it cannot be easily untied. The butterfly knot can be used to pull a rope bridge taut. This knot can be used to tighten a fixed rope when mechanical means are not available. It will not jam if a stick is placed between the two upper loops.

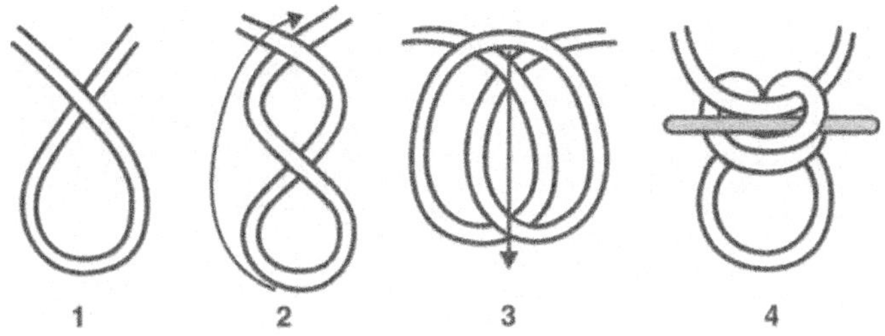

Table of Contents

Chapter 3. Weapons Handling

Chapter 6. Use of Cover and Concealment

Chapter 7. Rifle Presentation

Chapter 8. Effects of Weather

Chapter 9. Zeroing

Appendices

Chapter 1. Introduction to Rifle Marksmanship

All Marines share a common warfighting belief: "Every Marine a rifleman." This simple credo reinforces the belief that all Marines are forged from a common experience, share a common set of values, and are trained as members of an expeditionary force in readiness. As such, there are no "rear area" Marines, and no one is very far from the fighting during expeditionary operations. The Marine rifleman of the next conflict will be as in past conflicts: among the first to confront the enemy and the last to hang his weapon in the rack after the conflict is won.

1001. Role of the Marine Rifleman

Marine Corps forces are employed across the entire range of military operations. At one end is war, which is characterized by large-scale, sustained combat operations. At the other end of the scale are those actions referred to as military operations other than war (MOOTW). MOOTW focuses on deterring aggression, resolving conflict, promoting peace, and supporting civil authorities. These operations can occur before, during, and after combat operations. Training and preparation for MOOTW should not detract from the Corps' primary mission of training Marines to fight and win in combat. MOOTW normally does not involve combat. However, Marines always need to be prepared to protect themselves and respond to changing threats and unexpected situations. Whenever the situation warrants the application of deadly force, the Marine rifleman must be able to deliver well aimed shots to eliminate the threat. Sometimes the need for a well aimed shot may even be heightened by the presence of noncombatants in close proximity to the target. The proficient rifleman handles this challenge without unnecessarily escalating the level of violence or causing unnecessary collateral damage. The Marine rifleman must have the versatility, flexibility, and skills to deal with a situation at any level of intensity across the entire range of military operations.

To be combat ready, the Marine must be skilled in the techniques and procedures of rifle marksmanship and take proper care of his rifle. Even when equipped with the best rifle in the world, a unit with poorly trained riflemen cannot be depended upon to accomplish their mission. Usually, poorly trained riflemen either fail to fire their weapon or they waste ammunition by firing ineffectively. To send Marines into harm's way without thorough training in the use of their individual weapons carries undue risks for every Marine in the unit. On the other hand, well trained riflemen can deliver accurate fire against the enemy under the most adverse conditions. A well trained rifleman is not only confident that he can help his unit accomplish it's mission, he is confident that he can protect his fellow Marines and himself.

1002. Conditions Affecting Marksmanship in Combat

Many factors affect the application of marksmanship in combat; among them are—

- Most targets are linear in nature and will consist of a number of men or objects irregularly spaced along covered or concealed areas.
- Most targets can be detected by smoke, flash, dust, noise or movement, but will only be visible for a brief moment before taking cover.
- The nature of the target, irregularities of terrain and vegetation will often require a rifleman to use a position other than prone in order to fire effectively on the target.
- The time in which a target can be engaged is often fleeting.

1003. Combat Mindset

In a combat environment, the Marine must be constantly prepared for possible target engagement. When a target presents itself in combat, there may be very little time to take action. A Marine must be able to engage the target quickly and accurately. The unique demands of combat (i.e., stress, uncertainty) dictate that the Marine be both physically and mentally prepared to engage enemy targets. It will not be enough to simply know marksmanship techniques and procedures. The Marine must develop the mental discipline to prepare for enemy contact. In the confusion, noise, and stress of the combat environment, the Marine must have the ability to eliminate any hesitation, fear or uncertainty of action and to focus on the actions required to fire well-aimed shots. This is accomplished through establishment of a combat mindset. The key factors in the development of a combat mindset include both physical and mental preparation.

Physical Preparation

In combat, targets can present themselves without warning. Therefore, it is essential for the Marine to maintain proper balance and control of his weapon at all times so he can quickly assume a firing position, present the weapon, and accurately engage the target. However, speed alone does not equate to effective target engagement. The Marine should fire only as fast as he can fire accurately, never exceeding his physical ability to apply the fundamentals of marksmanship. To be effective in combat, the Marine must train to perfect the physical skills of shooting so those skills become second nature. Mastery of physical skills allow the Marine to concentrate on the mental aspects of target engagement; e.g., scanning for targets, detection of targets, selection and use of cover. The more physical skills that a Marine can perform automatically, the more concentration he can give to the mental side of target engagement.

Mental Preparation

While combat is unpredictable and constantly changing, the Marine can prepare himself mentally for confrontation with the enemy. The stress of battle, coupled with the often limited time available to engage targets, requires concentration on the mental aspects of target engagement; e.g., scanning for targets, detection of targets, and the selection and use of cover.

Knowledge of the Combat Environment

The Marine must be constantly aware of the surroundings to include the terrain, available cover, possible areas of enemy contact, backdrop of the target, etc. This awareness will enable the Marine to select and assume a firing position and to quickly and accurately engage targets.

Plan of Action

In combat, the situation will dictate the action to be taken. The Marine must understand the situation, identify and evaluate possible courses of action, and then develop a plan for target engagement that accomplishes the mission.

Confidence

The Marine must believe in his ability to engage targets accurately in any combat situation. A Marine's level of confidence is rooted in the belief that future challenges will be overcome—particularly the challenge of firing well aimed shots in a combat environment where the enemy may be returning fire. A key factor in a Marine's level of confidence is the degree to which he has mastered the techniques and procedures of the rifle marksmanship. Mastery of rifle marksmanship can only be obtained by quality instruction. Quality instruction is the foundation for practical application of the marksmanship fundamentals during range and field firing.

CHAPTER 2. INTRODUCTION TO THE M16A2 SERVICE RIFLE

Note

+The procedures in this manual are written for right-handed Marines; left-handed Marines should reverse instructions as necessary.

2001. DESCRIPTION

The M16A2 service rifle is a lightweight, 5.56 millimeter (mm), magazine-fed, gas-operated, air-cooled, shoulder-fired rifle. (Fig. 2-1 shows a right-side view and fig. 2-2, on page 2-2, shows the left-side view.) The rifle fires in either semiautomatic (single-shot) mode or a three-round burst through the use of a selector lever. The M16A2 service rifle has a maximum effective range of 550 meters for individual or point targets. The bore and chamber are chrome-plated to reduce wear and fouling. The handguards are aluminum-lined and are vented to permit air to circulate around the barrel for cooling purposes and to protect the gas tube. An aluminum receiver helps reduce the overall weight of the rifle. The trigger guard is equipped with a spring-loaded retaining pin that, when depressed, allows the trigger guard to be rotated out of the way for access to the trigger while wearing heavy gloves. An ejection port cover prevents dirt and sand from getting into the rifle through the ejection port. The ejection port cover should be closed when the rifle is not being fired. It is automatically opened by the action of the bolt carrier. The muzzle compensator serves as a flash suppressor and assists in reducing muzzle climb.

2002. OPERATIONAL CONTROLS

Selector Lever

The selector lever has three settings—safe, semi, and burst. The setting selected depends on the firing situation. See figure 2-3 on page 2-2.

Safe

The selector lever in the safe position prevents the rifle from firing.

Semi

The selector lever in the semi position allows one shot to be fired with each pull of the trigger.

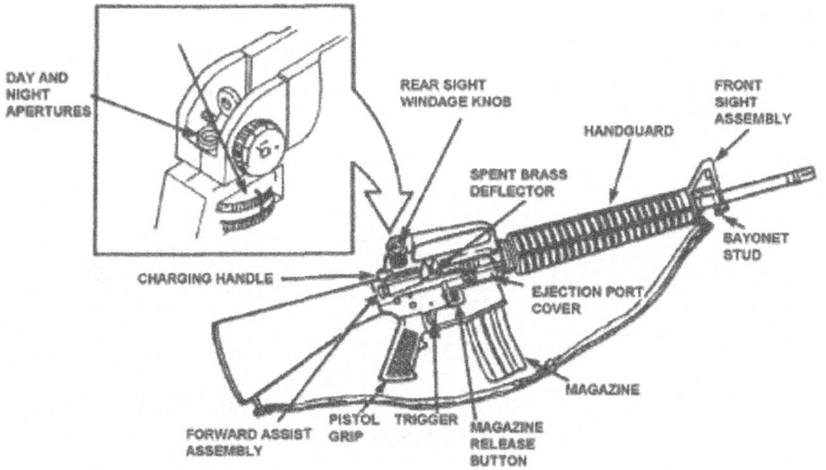

Figure 2-1. M16A2 Service Rifle (Right Side View).

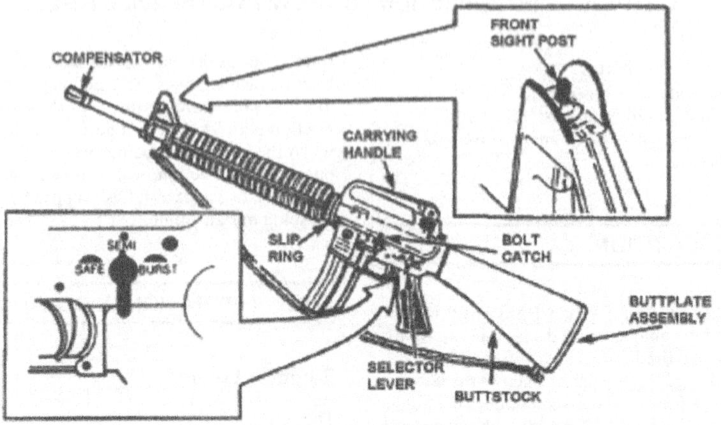

Figure 2-2. M16A2 Service Rifle (Left Side View).

Burst

The selector lever in the burst position allows the rifle to continue its cycle of operation until interrupted by the burst cam. With each pull of the trigger, the burst cam limits the maximum number of rounds fired to three. The burst cam is not "self-indexing." If burst is selected, the burst cam does not automatically reset to the first shot position of the three-round burst. One, two or three shots may be fired on the first pull of the trigger. Each subsequent pull of the trigger results in a complete three-round burst unless the trigger is released before the cycle is complete. If the trigger is released during the burst and the three-round cycle is interrupted, the next pull of the trigger fires the rounds remaining in the interrupted three-round cycle.

Magazine Release Button

The magazine release button releases the magazine from the magazine well. See figure 2-4.

Charging Handle

When the charging handle is pulled to the rear, the bolt unlocks from the barrel extension locking lugs and the bolt carrier group moves to the rear of the receiver. See figure 2-5.

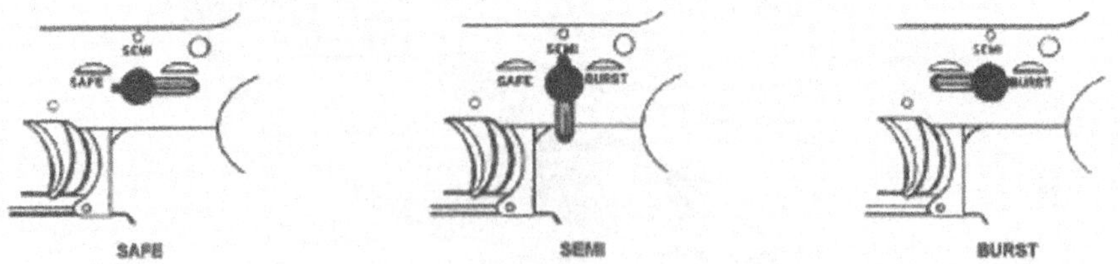

Figure 2-3. Selector Lever.

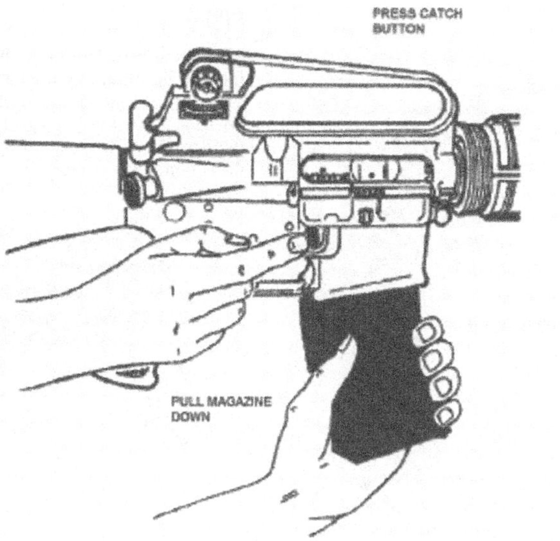

Figure 2-4. Magazine Release Button.

Bolt Catch

If the charging handle is pulled to the rear when the lower portion of the bolt catch is depressed, the bolt carrier group will lock to the rear. When the bolt carrier group is locked to the rear and the upper portion of the bolt catch is depressed, the bolt carrier group will slide forward, driven by the buffer assembly and action spring, into the firing position.

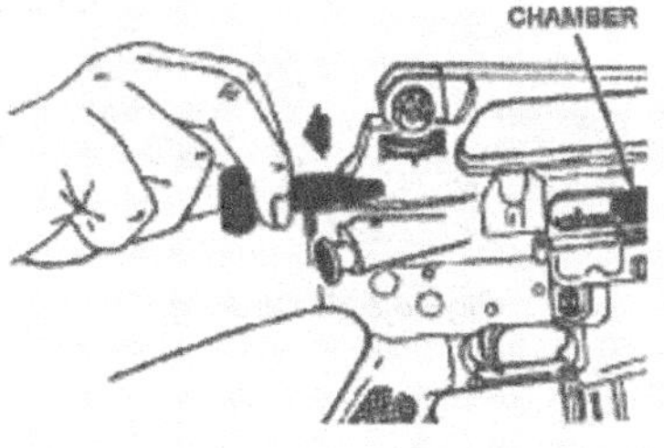

Figure 2-5. Charging Handle.

2003. CYCLE OF OPERATION

Firing

The hammer releases and strikes the head of the firing pin, driving the firing pin into the round's primer. The primer ignites the powder in the cartridge. Gas generated by the rapid burning of powder propels the projectile through the barrel. After the projectile passes the gas port, a portion of the expanding gas enters the gas port and gas tube. The gas tube directs the gas rearward into the bolt carrier key and causes the bolt carrier to move rearward. See figure 2-6 on page 2-4.

Unlocking

Figure 2-7 on page 2-4 illustrates unlocking of the bolt. As the bolt carrier moves to the rear, the bolt cam pin follows the path of the cam track located in the bolt carrier. This causes the bolt assembly to rotate until the bolt-locking lugs are no longer aligned behind the barrel extension locking lugs.

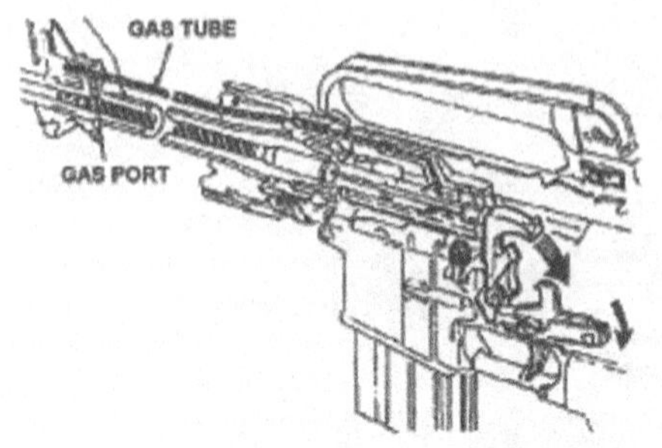

Figure 2-6. Firing.

Figure 2-7. Unlocking.

Extracting

As the bolt carrier group continues to move to the rear, the extractor claw withdraws the cartridge case from the chamber. See figure 2-8.

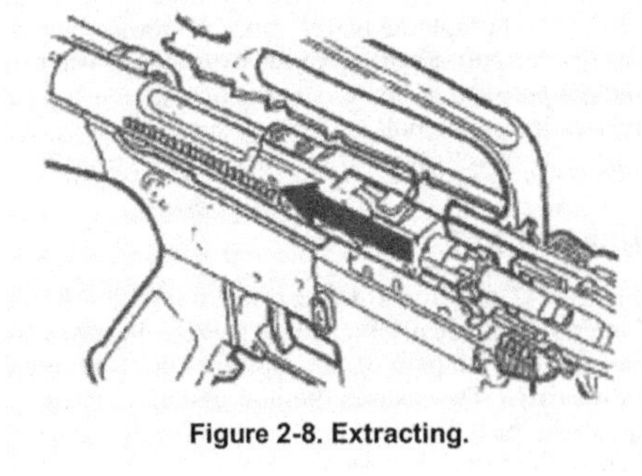

Figure 2-8. Extracting.

Ejecting

The ejector, located in the bolt face, is compressed into the bolt body by the base of the cartridge case. The rearward movement of the bolt carrier group allows the nose of the cartridge case to clear the front of the ejection port. The cartridge case is thrown out by the action of the ejector and spring. See figure 2-9.

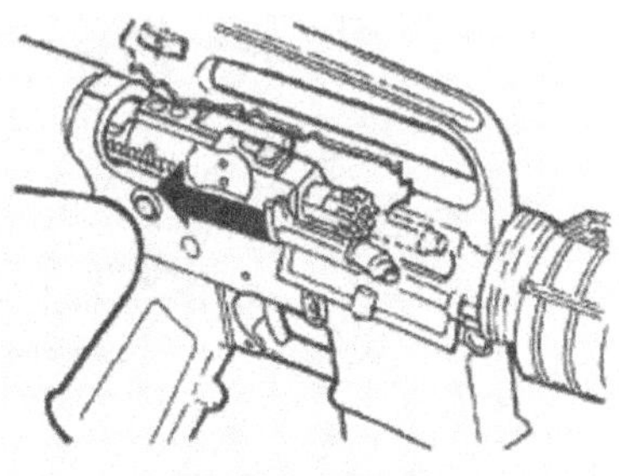

Figure 2-9. Ejecting.

Cocking

Continuing its rearward travel, the bolt carrier overrides the hammer, forces it down into the receiver, compresses the hammer spring, and causes the disconnector to engage the lower hammer hook. See figure 2-10.

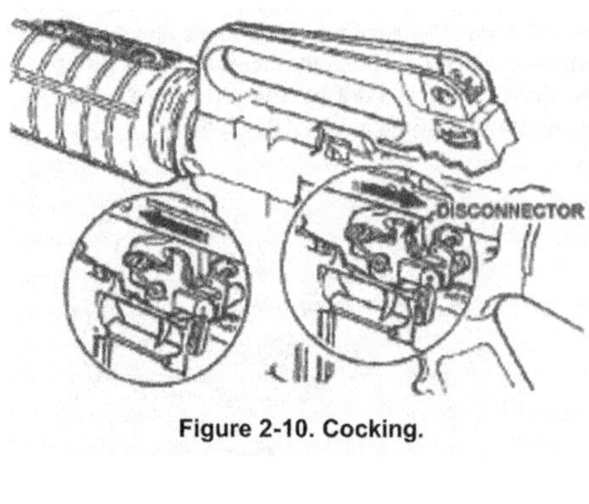

Figure 2-10. Cocking.

Feeding

Once rearward motion causes the bolt carrier group to clear the top of the magazine, the expansion of the magazine spring forces a round into the path of the bolt. After the action spring overcomes and absorbs the rearward motion of the bolt carrier group, it expands and sends the buffer assembly and bolt carrier group forward with enough force to strip a round from the magazine. See figure 2-11.

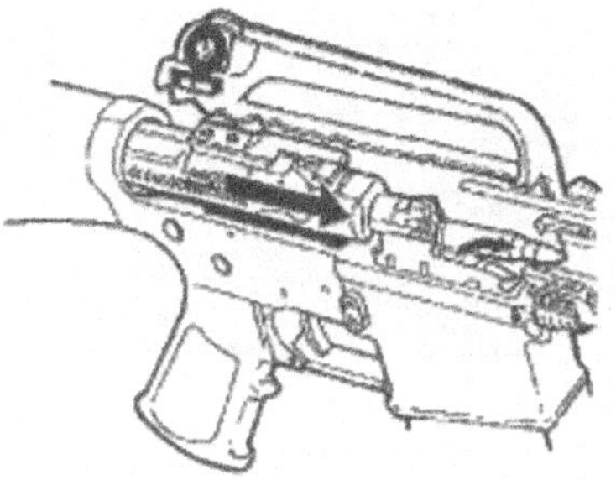

Figure 2-11. Feeding.

Chambering

As the bolt carrier group continues to move forward, pushing a fresh round in front of it, the face of the bolt thrusts the new round into the chamber. The extractor claw grips the rim of the cartridge case. The ejector is forced into its hole, compressing the ejector spring. See figure 2-12.

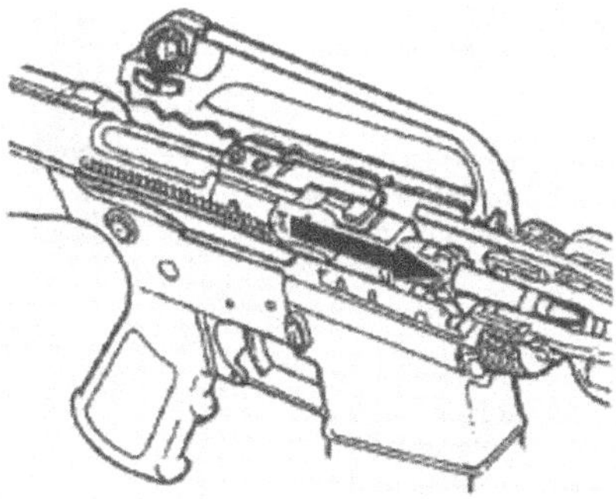

Figure 2-12. Chambering.

Locking

As the bolt carrier group continues to move forward, the bolt-locking lugs are forced against the barrel extension and the bolt cam pin is forced along the cam track. The bolt rotates and aligns the bolt locking lugs behind the barrel extension locking lugs. The weapon is ready to fire. See figure 2-13.

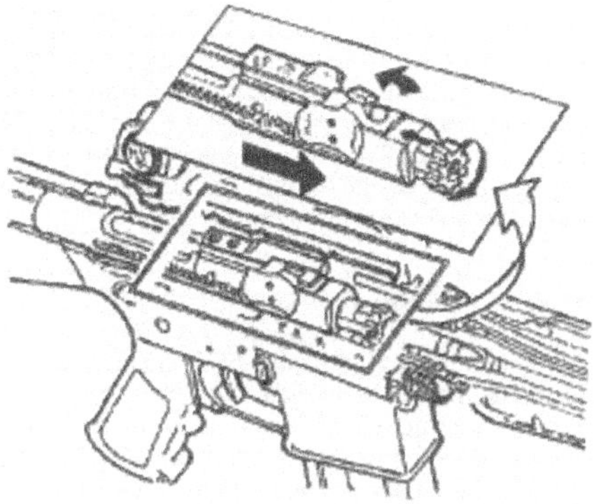

Figure 2-13. Locking.

2004. Ammunition

Four types of ammunition are authorized for use with the M16A2 service rifle: ball (M193 and M855), tracer (M196 and M856), dummy (M199), and blank (M200) (see fig. 2-14).

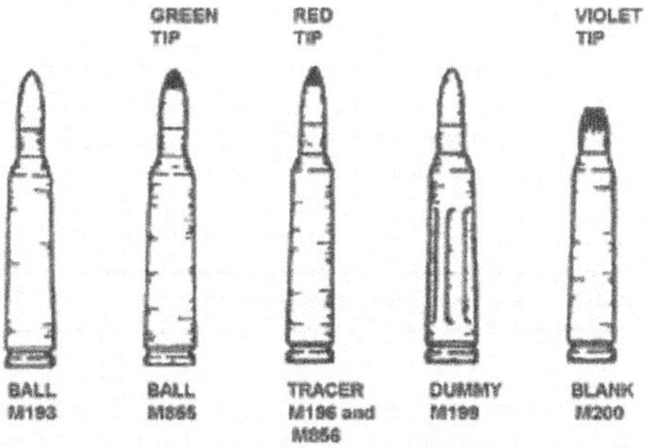

Figure 2-14. Authorized Ammunition.

M193 Ball

This ammunition is a 5.56mm centerfire cartridge with a 55-grain gilded-metal jacket, lead alloy core bullet. The primer and case are waterproofed. The M193 ball ammunition has no identifying marks.

M855 Ball

This ammunition is the primary ammunition for the M16A2 rifle. Identified by a green tip, its 5.56mm centerfire cartridge has better penetration than the M193. It has a 62-grain gilded-metal jacket bullet. The rear two-thirds of the core of the projectile is lead alloy and the front one-third is a solid steel penetrator. The primer and case are waterproofed.

M196 and M856 Tracer

This ammunition has the same basic characteristics as ball ammunition. Identified by a bright red tip, its primary uses include observation firing, incendiary effect, and signaling. Tracer ammunition should be intermixed with ball ammunition in a ratio no greater than 1:1. The preferred ratio is one tracer to four balls (1:4) to prevent metal fouling in the bore.

M199 Dummy

This ammunition has six grooves along the side of the case. It contains no propellants or primer. The primer well is open to prevent damage to the firing pin. The dummy cartridge is used during dry fire and other training purposes.

M200 Blank

This ammunition has the case mouth closed with a seven-petal rosette crimp. It contains no projectile. Blank ammunition, identified by its violet tip, is used for training purposes.

2005. Preventive Maintenance

Normal care and cleaning of the rifle will result in proper functioning of its all parts. Only issue-type cleaning materials maybe used. Improper maintenance can cause stoppages, reducing combat readiness and effectiveness.

Main Group Disassembly

The M16A2 service rifle is disassembled into three main groups (see fig. 2-15). Before disassembling the M16A2 service rifle—

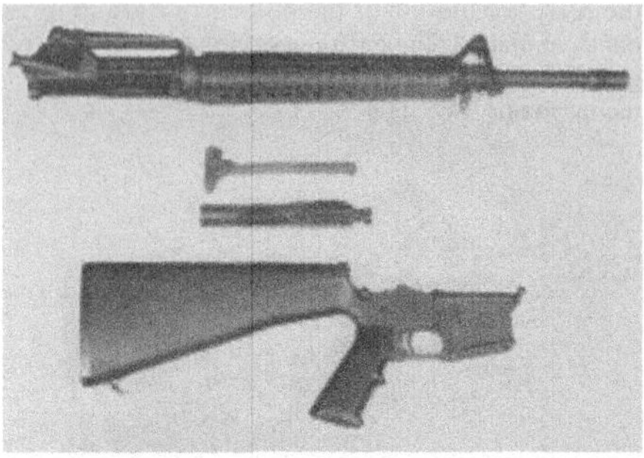

Figure 2-15. Three Main Groups.

- Ensure the rifle is in Condition 4 (see para. 3002).
- Remove the sling.

Note

To facilitate control and ease of disassembly, the handguards may be removed using the "buddy system" as described in Technical Manual (TM) 05538C-10/1A, *Operator's Manual w/Components List Rifle, 5.56-mm, M16A2E W/E (1005-01-128-9936).*

Upper Receiver

To disassemble the upper receiver—

- Place the rifle on the buttstock. Press down on the slip ring with both hands. Pull the handguards free (see fig. 2-16).

Use caution when the handguards are off the rifle. Handguards provide protection for the gas tube. Damage to the gas tube adversely affects the functioning of the rifle.

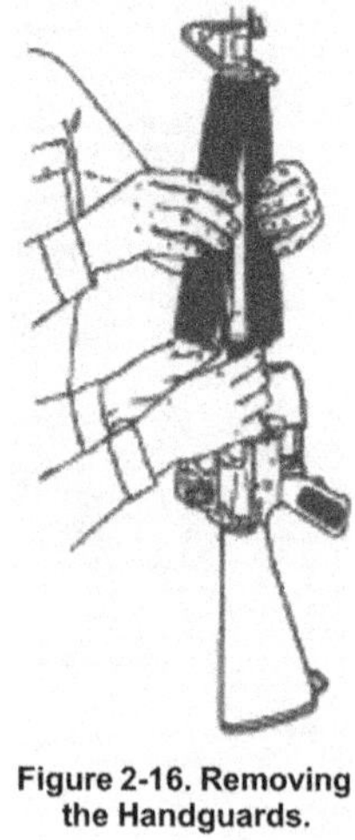

Figure 2-16. Removing the Handguards.

- Move the take down pin from the left to the right as far as it will go to allow the lower receiver to pivot down from the upper receiver.
- Move the receiver pivot pin from left to right as far as it will go and separate the upper and lower receivers.
- Pull back the charging handle and bolt carrier about 3 inches and remove the bolt carrier group.
- Remove the charging handle by sliding it back and down, out of the upper receiver.

No further disassembly is conducted on the upper receiver group.

Bolt Carrier

To disassemble the bolt carrier—

- Remove the firing pin retaining pin.
- Push the bolt back into the bolt carrier to the locked position.
- Tap the base of the bolt carrier against the palm of your hand so the firing pin will drop out.
- Rotate the bolt cam pin one-quarter turn and lift the bolt cam pin out.
- Withdraw the bolt assembly from the carrier.
- Press on the extractor's rear and use the firing pin to push out the extractor-retaining pin. Remove the extractor and spring (the spring is permanently attached to the extractor). (See fig. 2-17.)

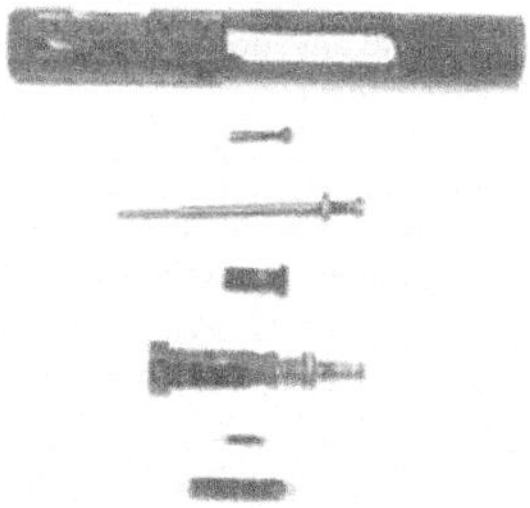

Figure 2-17. Bolt Carrier Disassembled.

CAUTION

Be careful not to damage the tip of the firing pin while pushing out the extractor-retaining pin.

Note

The extractor assembly has a rubber insert within the spring. Do not attempt to remove it. If the spring comes loose, put the large end of the spring in the extractor and seat it. Push in the extractor pin.

Lower Receiver

To disassemble the lower receiver—

- Press in the buffer and depress the buffer retainer.

Note

It may be necessary to use the edge of the charging handle to depress the buffer retainer.

- Press the hammer downward and ease the buffer and action spring forward and out of the receiver.
- Separate the parts. See figure 2-18.

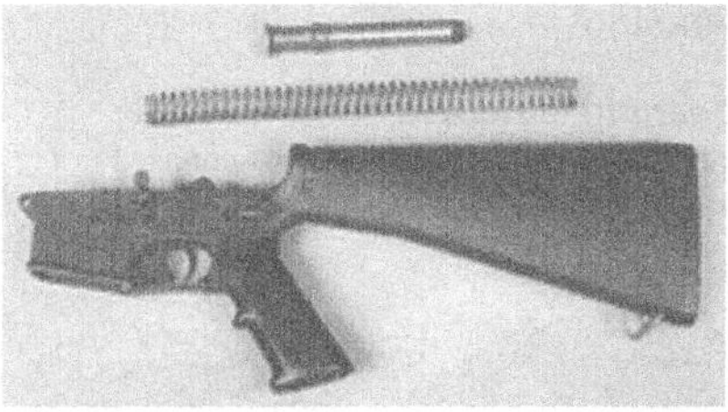

Figure 2-18. Lower Receiver Disassembled.

No further disassembly of the lower receiver is performed.

Note

In combat situations, the rifle may be partially disassembled in any sequence. However, combat situations are the exception, not the rule. Under normal operating circumstances, disassemble the rifle in the sequence just performed. Any further disassembly of the rifle is to be performed by a qualified armorer.

Magazine Disassembly

The magazine should be disassembled regularly for cleaning to avoid the possibility of malfunction or stoppage of the rifle caused by dirty or damaged magazines. To disassemble the magazine—

- Pry up and push base plate out from the magazine.
- Jiggle the spring and follower to remove. Do not remove the follower from the spring. See figure 2-19.

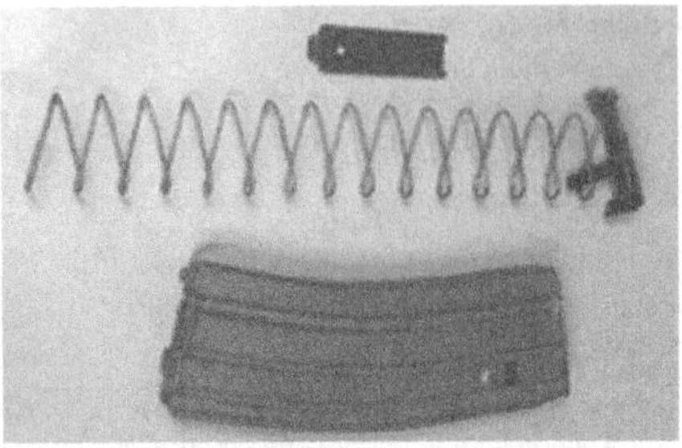

Figure 2-19. Magazine Disassembled.

Cleaning

Cleaning Materials

The following cleaning materials are used in preventive maintenance (see fig. 2-20):

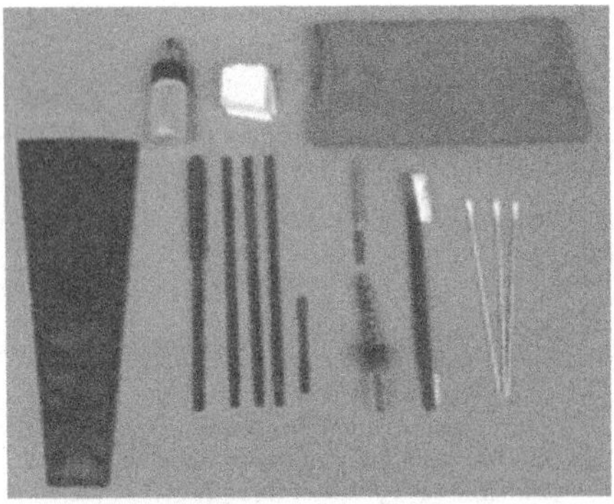

Figure 2-20. Cleaning Materials.

- Cleaner, lubricant, and preservative (CLP). Always shake the bottle well before use.
- Rod in three sections and a handle assembly.
- Patch holder section, swabs, patches, pipe cleaners, and clean rags.
- Brushes: bore, chamber, and general purpose.

Cleaning the Upper Receiver

Basic cleaning of the upper receiver group should include the following:

- Attach the three rod sections together but leave each one about two turns short of being tight.
- Attach the patch holder onto the rod.
- Point the muzzle down and insert the non-patch end of the rod into the chamber. Attach the handle to the cleaning rod section and pull a CLP-moistened 5.56mm patch through the bore.
- Attach the bore brush to the rod but leave it two turns short of being tight. Put a few drops of CLP on the bore brush. Insert the rod into the barrel from the chamber end, attach the handle, and pull the brush through the bore. Repeat 3 times. Remove bore brush and attach the patch holder to the rod with a CLP moistened patch insert the rod into the barrel from the chamber end, attach the handle, and pull the patch through the bore.
- Inspect the bore for cleanliness by holding the muzzle to your eye and looking into the bore.
- Repeat the above steps until the patches come out of the bore clean.
- Attach the chamber brush and one section of the cleaning rod to the handle. Moisten it well with CLP and insert it into the chamber.
- Scrub the chamber and bolt lugs using a combination of a plunging and clockwise rotating action.

Note

Do not reverse direction of the brush while it is in the chamber.

- Clean the interior portion of the upper receiver with the general-purpose brush and CLP.
- Dry the bore, chamber, and the interior of the receiver with rifle patches, swabs, and clean rags until they come out clean. Then moisten all interior surfaces with CLP.
- Wipe the barrel, gas tube, and handguards clean with a rag.

Cleaning the Bolt Carrier Group

- Clean the outer and inner surfaces of the bolt carrier with a general-purpose brush.
- Clean the bolt carrier key with a pipe cleaner.
- Clean the locking lugs, gas rings, and exterior of the bolt with the general-purpose brush.
- Insert a swab into the rear of the bolt and swab out the firing pin recess and gas ports.

- Clean the extractor with the general-purpose brush, ensuring all the carbon is removed from underneath the extractor lip.
- Clean extractor pin, firing pin, and firing pin retaining pin using the general-purpose brush and CLP.
- Clean charging handle assembly with the general-purpose brush and patches.

Cleaning the Lower Receiver

- Wipe dirt from the firing mechanism using a general-purpose brush, clean patch, pipe cleaners, and swabs.
- Clean the outside of the receiver with the general-purpose brush and CLP. Clean the buttplate and rear sling swivel, ensuring drain hole is clear of dirt.
- Wipe the inside of the buffer tube, buffer, and action spring.
- Wipe the inside of the magazine well with a rag.
- Wipe out the inside of the pistol grip and ensure that it is clean.

Cleaning the Magazine

- Clean the inside of the magazine with the general-purpose brush and CLP.
- Wipe dry.
- Keep the spring lightly oiled.

Inspection

While cleaning the rifle, and during each succeeding step in the preventive maintenance process, inspect each part for cracks and chips and ensure parts are not bent or badly worn. Report any damaged part to the armorer. Inspection is a critical step to ensure the combat readiness of your rifle. It is performed normally during rifle cleaning (prior to lubrication), however, it can be performed throughout the preventive maintenance process.

Lubrication

Lubrication is performed as part of the detailed procedure for preventive maintenance. Lubrication procedures are also performed in preparation for firing.

Lubricant

In all but the coldest arctic conditions, CLP is the lubricant for the rifle. Remember to remove excess CLP from the bore and chamber before firing.

- Lightly lube means that a film of CLP barely visible to the eye should be applied.
- Generously lube means the CLP should be applied heavily enough that it can be spread with the finger.

Upper Receiver

- Lightly lube the inside of the upper receiver, bore, chamber, outer surfaces of the barrel, and surfaces under the handguard.
- Depress the front sight detent and apply two or three drops of CLP to the front sight detent. Depress several times to work lubrication into the spring.
- Lubricate the moving parts and elevation screw shaft of the rear sight.

Bolt Carrier Group

- Generously lube the outside of the cam pin area, the bolt rings, and outside the bolt body.
- Lightly lube the charging handle and the inner and outer surfaces of the bolt carrier.

Lower Receiver

- Lightly lube the inside of the lower receiver extension.
- Generously lube the moving parts inside the lower receiver and their pins.

Reassembly

Reassembling the Rifle

- Return all cleaning gear into the buttstock of the rifle and close the buttplate.
- Connect the buffer and action spring and insert them into the buffer tube/stock.
- Place the extractor and spring back on the bolt. Depress the extractor to align the holes and reinsert the extractor pin.
- Insert the bolt into the carrier. Do not switch bolts between rifles.
- Hold the bolt carrier with the bolt carrier key at 12 o'clock. Insert the bolt into the bolt carrier with the extractor at 12 o'clock.

- Rotate the bolt counterclockwise until the cam pinhole aligns to the cam pin slot in the bolt carrier.

WARNING WARNING WARNING WARNING

WARNING Ensure the cam pin is installed in the bolt group or the rifle may explode while firing.

- Insert the bolt cam pin through the bolt carrier and into the bolt. Rotate the cam pin 1/4 turn right or left. Pull the bolt forward until it stops.
- Drop in the firing pin from the rear of the bolt carrier and seat it.
- Replace the firing pin retaining pin. Ensure the head of the firing pin retaining pin is recessed inside the bolt carrier. The firing pin should not fall out when the bolt carrier group is turned upside down.
- Place the charging handle in the upper receiver by lining it up with the grooves in the receiver. Push the charging handle partially in.
- With the bolt in the unlocked position, place bolt carrier key into the groove of the charging handle.
- Push the charging handle and bolt carrier group into the upper receiver until the charging handle locks.
- Join the upper and lower receivers and engage the receiver pivot pin.
- Ensure the selector lever is on safe before closing the upper receiver. Close the upper and lower receiver groups. Push in the takedown pin.
- Install the handguards.
- Attach the sling.

Reassembling the Magazine

To reassemble the magazine—

- Insert the follower and jiggle the spring to install.
- Slide the base under all four tabs until the base catches. Make sure the printing is on the outside.

2006. Function Check

A function check is performed to ensure the rifle operates properly. To perform a function check:

- Place the weapon in Condition 4 (see para. 3002).
- Pull the charging handle to the rear and release. Ensure the selector lever is on safe and pull the trigger. The hammer should not fall.
- Place the selector lever on semi. Pull the trigger and hold it to the rear. The hammer should fall. Pull the charging handle to the rear and release. Release the trigger and pull again. The hammer should fall.
- Pull the charging handle to the rear and release. Place the selector lever on burst. Pull the trigger and hold it to the rear. The hammer should fall. Pull the charging handle to the rear three times and release. Release the trigger and pull again. The hammer should fall.
- Pull the charging handle to the rear and release. Place the selector lever on safe.

2007. User Serviceability Inspection

Individual Marines must perform user serviceability inspections on their weapons before firing them. This inspection ensures the weapon is in an acceptable operating condition.

- Place rifle in Condition 4 (see para. 3002).
- Conduct a function check.
- Check the rifle to ensure the following:
- Compensator is tight.
- Barrel is tight.
- Front sight post is straight.
- Front sight post is adjustable.
- Handguards are serviceable.
- Rear sight elevation and windage knobs are adjustable and have distinct clicks.
- Stock is tight on the lower receiver.
- Weapon is properly lubricated for operational conditions.
- Barrel is clear of obstructions.
- Gas rings are serviceable.
- Ensure magazines are serviceable.
- Load the rifle with an empty magazine. Ensure that the magazine can be seated.
- Without depressing the bolt catch, pull the charging handle to the rear. Ensure that the bolt locks to the rear.
- Depress the upper portion of the bolt catch and observe bolt moving forward on an empty chamber.

Ensure the bolt moves completely forward and locks in the chamber.

- Repeat this procedure with all magazines.

2008. Field Maintenance

Preventive maintenance in the field is performed when detailed disassembly and cleaning is not practical due to operational tempo or the level of threat. To perform limited field preventive maintenance—

- Place the rifle in Condition 4 (see paragraph 3002).
- Break the rifle down by removing the rear take down pin and rotating the upper receiver and barrel forward.
- Remove the bolt carrier group.
- Do not disassemble the bolt carrier group further.
- Clean the bolt carrier group.
- Clean the upper and lower receiver groups (without further disassembly).
- Clean the bore and chamber.
- Lubricate the rifle.
- Reassemble the rifle and perform a user serviceability inspection.

2009. Cleaning the Rifle in Various Conditions

The climatic conditions in various locations require special knowledge about cleaning and maintaining the rifle. The conditions that will affect the rifle the most are: hot, wet tropical; hot, dry desert; arctic or low temperature; and heavy rain and fording.

Hot, Wet Tropical

- Perform normal maintenance.
- Clean and lubricate your rifle more often. Inspect hidden surfaces for corrosion. Pay particular attention to spring-loaded detentes.
- Use lubricant more liberally.
- Unload and check the inside of the magazine more frequently. Wipe dry and check for corrosion.
- When practical, keep the rifle covered.

Hot, Dry Desert

Hot dry climates are usually areas that contain blowing sand and fine dust. Dust and sand will get into the rifle and magazines, causing stoppages. It is imperative to pay particular attention to the cleaning and lubrication of the rifle in this type of climate.

Corrosion is less likely to form in these environments, and lubrication will attract more dirt. For this reason, use lubrication more sparingly.

Whenever practical, keep the rifle covered.

Arctic or Low Temperature

- Clean and lubricate the rifle in a warm room, with the rifle at room temperature, if possible. Lubricating oil, arctic weapons (LAW) can be used below a temperature of 0 degrees Fahrenheit and must be used below -35 degrees Fahrenheit.
- Keep the rifle covered when moving from a warm to a cold environment to allow gradual cooling of the rifle. This prevents the condensation of moisture and freezing. Condensation will form on the rifle when it is moved from outdoors to indoors. If possible, leave the rifle in a protected but cold area outdoors. When bringing the rifle inside to a warm place, it should be disassembled and wiped down several times as it warms.
- Always try to keep the rifle dry.
- Unload and perform a function check every 30 minutes, if possible, to help prevent freezing of functional parts.
- Do not lay a warm rifle in snow or ice.
- Keep the inside of the magazine and ammunition wiped dry. Moisture will freeze and cause stoppages.

Heavy Rain and Fording

- Keep the rifle dry and covered when practical.
- Keep water out of the barrel if possible. If water does get in, drain and (if possible) dry with a patch. If water is in the barrel, point the muzzle down and break the seal by doing a chamber check so the water will drain. If water is in the stock of the weapon, ensure the drain hole in the stock is clear so the water can run out.
- Perform normal maintenance.

CHAPTER 3. WEAPONS HANDLING

Weapons handling procedures provide a consistent and standardized way for a Marine to handle, operate, and employ the rifle safely and effectively. Proper weapons handling procedures ensure the safety of Marines by eliminating negligent discharges and reinforcing positive identification of targets before engagement. Weapons handling procedures apply at all levels of training and during combat operations.

Note

+The procedures in this manual are written for right-handed Marines; left-handed Marines should reverse instructions as necessary.

3001. Safety Rules

Safe handling of the rifle is critical. If proper weapons handling procedures are not used, a Marine risks his safety and the safety of his fellow Marines. During combat, a Marine must react quickly, safely, and be mentally prepared to engage targets. To ensure that only the intended target is engaged, a Marine must apply the following safety rules at all times:

Rule 1—Treat every weapon as if it were loaded. When a Marine takes charge of a rifle in any situation, he must treat the weapon as if it were loaded, determine its condition (see para. 3003), and continue applying the other safety rules.

Rule 2—Never point a weapon at anything you do not intend to shoot. A Marine must maintain muzzle awareness at all times.

Rule 3—Keep your finger straight and off the trigger until you are ready to fire. A target must be identified before taking the weapon off safe and moving the finger to the trigger.

Rule 4—Keep the weapon on safe until you intend to fire. A target must be identified before taking the weapon off safe. This rule is intended to eliminate the chance of the weapon discharging by accident (e.g., brush snagging the trigger).

3002. Weapons Conditions

A weapon's readiness is described by one of fou conditions. The steps in the loading and unloadin process take the rifle through four specific condition of readiness for live fire.

Condition 1. Safety on, magazine inserted, roun in chamber, bolt forward, ejectio port cover closed.

Condition 2. Not applicable to the M16A2 rifle.

Condition 3. Safety on, magazine inserted, cham ber empty, bolt forward, ejectio port cover closed.

Condition 4. Safety on, magazine removed, cham ber empty, bolt forward, ejectio port cover closed.

3003. Determining a Weapon's Condition (Chamber Check)

A Marine must know the condition of his weapon a all times. When a Marine takes charge of a weapon i any situation, he must determine its condition Situations include coming across an unmanned rifle i combat, taking charge of any weapon after it has bee unmanned (e.g., out of a rifle rack, stored in a vehicle) or taking charge of another Marine's weapon. T determine the condition of the weapon in any of thes situations, the Marine should:

- Determine if a magazine is present.
- Ensure the rifle is on safe.

- Conduct a chamber check.
- Bring the left hand back against the magazine well.
- Extend the fingers of the left hand and cover the ejection port (see fig. 3-1).

Figure 3-1. Position of Hand.

- Grasp the charging handle with the index and middle fingers of the right hand.
- Pull the charging handle slightly to the rear and visually and physically inspect the chamber (see fig. 3-2). *Right-handed Marines*, insert one finger of the left hand into the ejection port and feel whether a round is present. *Left-handed Marines*, insert the thumb of the right hand into the ejection port and feel whether a round is present.

Figure 3-2. Chamber Check.

Note

The same procedure is used in daylight as during low visibility. A chamber check may be conducted at any time.

CAUTION

Pulling the charging handle too far to the rear while inspecting the chamber may cause double feed or ejection of one round of ammunition.

- Release the charging handle and observe the bolt going forward.
- Tap the forward assist.
- Close the ejection port cover (if time and the situation permit).
- Remove the magazine (if present) and observe if ammunition is present. If time permits, count the rounds. Reinsert the magazine into magazine well.

3004. Weapons Commands

Weapons commands dictate the specific steps required to load and unload the rifle. Six commands are used in weapons handling:

Load. This command is used to take the weapon from Condition 4 to Condition 3.

Make Ready. This command is used to take the weapon from Condition 3 to Condition 1.

Fire. This command is used to specify when a Marine may engage targets.

Cease-Fire. This command is used to specify when a Marine must stop target engagement.

Unload. This command is used to take the weapon from any condition to Condition 4.

Unload and Show Clear. This command is used when an observer must check the weapon to verify that no ammunition is present before the rifle is placed in Condition 4.

Loading the Rifle

Perform the following steps to load the rifle (take the rifle to Condition 3):

- Ensure the rifle is on safe.
- Withdraw the magazine from the magazine pouch.
- Observe the magazine to ensure it is filled.
- Fully insert the magazine in the magazine well. Without releasing the magazine, tug downward on the magazine to ensure it is seated.
- Close the ejection port cover.
- Fasten the magazine pouch.

Making the Rifle Ready

Perform the following steps to make the rifle ready for firing (take the rifle to Condition 1):

- Pull the charging handle to the rear and release. There are two methods of doing this:
- Grip the pistol grip firmly with the right hand and pull the charging handle with the left hand to its rearmost position and release (see fig. 3-3).

  Or grip the handguards firmly with the left hand and pull the charging handle with the right hand to its rearmost position and release (see fig. 3-4).

Figure 3-3. Pulling the Charging Handle with the Left Hand.

Figure 3-4. Pulling the Charging Handle with the Right Hand.

- To ensure ammunition has been chambered, conduct a chamber check (see para. 3003) to ensure a round has been chambered.
- Check the sights (to ensure proper battlesight zero [BZO] setting, correct rear sight aperture, etc.).
- Close ejection port cover.

Fire

On the command "Fire," aim the rifle, take the rifle off safe, and pull the trigger.

Cease-Fire

On the command "Cease Fire," perform the following steps:

- Place your trigger finger straight along the receiver.
- Place the weapon on safe.

Unloading the Rifle

Perform the following steps to unload the rifle (take the rifle to Condition 4):

- Ensure the weapon is on safe.
- Remove the magazine from the rifle and retain it on your person.

- Cup the left hand under the ejection port, rotate the weapon until the ejection port faces down.
- Pull the charging handle to the rear and catch the round in the left hand (see fig. 3-5).

Figure 3-5. Catching the Round.

- Lock the bolt to the rear.
- Put the weapon on safe if the selector lever would not move to safe earlier.
- Ensure the chamber is empty and that no ammunition is present.
- Depress the bolt catch and observe the bolt moving forward on an empty chamber (see fig. 3-6).

Figure 3-6. Observing the Chamber.

- Close the ejection port cover.
- Check the sights (for proper BZO setting, correct rear sight aperture, etc.).
- Place any ejected round into the magazine and return the magazine to the magazine pouch and close the magazine pouch.

Unloading and Showing the Rifle Clear

Perform the following steps to unload the rifle and show it clear to an observer (take the rifle to Condition 4). The Marine—

- Ensures the weapon is on safe.
- Removes the magazine from the rifle and retains it.
- Cups the left hand under the ejection port, rotates the weapon until the ejection port faces down.
- Pulls the charging handle to the rear and catches the round in the left hand.
- Locks the bolt to the rear and ensures the chamber is empty and that no ammunition is present.
- Has another Marine inspect the weapon to ensure no ammunition is present (see fig. 3-7).

Figure 3-7. Observer Inspection.

The observer—

- Visually inspects the chamber to ensure it is empty, no ammunition is present, and the magazine is removed.
- Ensures the weapon is on safe.
- Acknowledges the rifle is clear.

The Marine, after receiving acknowledgment that the rifle is clear—

- Depresses the bolt catch and observes the bolt moving forward on an empty chamber.
- Closes the ejection port cover.
- Checks the sights (for proper BZO setting, correct rear sight aperture, etc.).
- Places any ejected round into the magazine and returns the magazine to the magazine pouch and closes the magazine pouch.

3005. Filling, Stowing, and Withdrawing Magazines

Filling the Magazine with Loose Rounds

Perform the following steps to fill the magazine:

- Remove a magazine from the magazine pouch.
- Place a round on top of the follower.
- Press down until the round is held between the follower and magazine feed lips (see fig. 3-8).

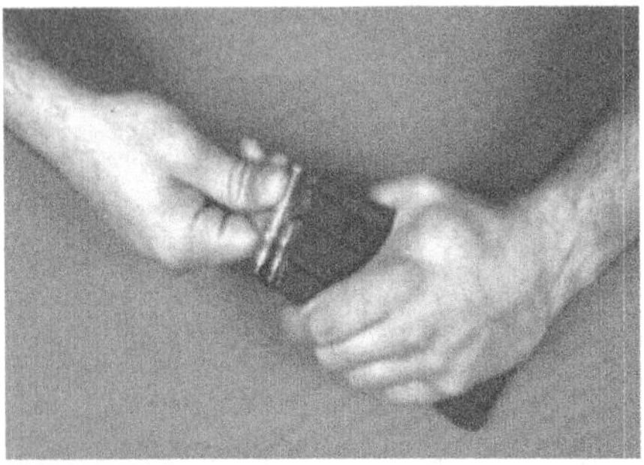

Figure 3-8. Filling the Magazine.

- Repeat until the desired number of rounds is inserted. The recommended number of rounds per magazine is 28 or 29. Thirty rounds in the magazine may prohibit the magazine from seating properly on a closed bolt.
- Tap the back of the magazine to ensure the rounds are seated against the back of the magazine.

Filling the Magazine Using a 10-round Stripper Clip and Magazine Filler

The magazine can also be filled quickly using a 10-round stripper clip and the magazine filler (see fig. 3-9).

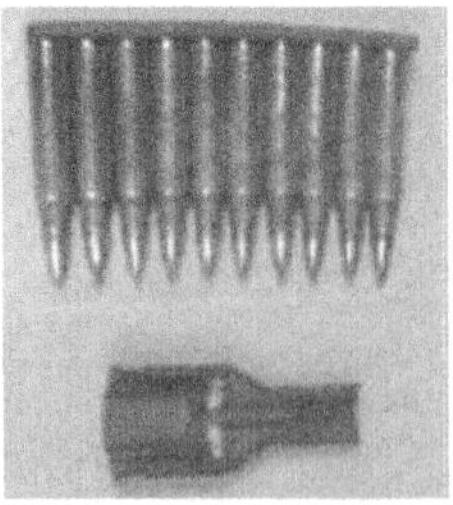

Figure 3-9. Magazine Filler and 10-round

Perform the following steps to fill the magazine with the 10-round stripper clip (see fig. 3-10):

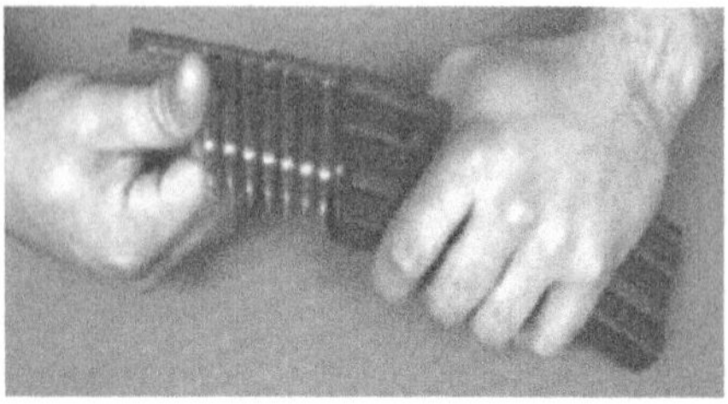

Figure 3-10. Filling the Magazine with a Stripper Clip and Magazine Filler.

- Remove a magazine from the magazine pouch.
- Slide the magazine filler into place.
- Place a 10-round stripper clip into the narrow portion of the magazine filler.
- Using thumb pressure on the rear of the top cartridge, press down firmly until all ten rounds are below the feed lips of the magazine.
- Remove the empty stripper clip while holding the magazine filler in place.
- Repeat until the desired number of rounds is inserted. The recommended number of rounds per magazine is 28 or 29. Thirty rounds in the magazine may prohibit the magazine from seating properly on a closed bolt.
- Remove magazine filler and retain it for future use.
- Tap the back of the magazine to ensure the rounds are seated against the back of the magazine.

Stowing Magazines

Magazine Pouch

In a magazine pouch, filled magazines are stored with rounds down and projectiles pointing away from the body.

Load-bearing Vest

In a load-bearing vest, filled magazines are stored with rounds down and projectiles pointing outboard.

Empty or Partially Filled Magazines

Empty or partially filled magazines are stored with the follower up to allow the selection of filled magazines by touch (i.e., at night).

Withdrawing Magazines

Magazine Pouch

With the right hand, withdraw magazines from the magazine pouch on the right side of the body. With the left hand, withdraw magazines from the magazine pouch on the left side of the body. To withdraw magazines from a magazine pouch—

- Use the thumb and index finger, pinch the magazine pouch release to open the magazine pouch.
- Slide the thumb over the magazines, feeling for a base plate indicating a filled magazine. Continue sliding the thumb until it rests on the back of the magazine.
- Grasp the magazine with the thumb, index finger, and middle finger and lift the magazine directly out of the pouch.
- Once the magazine is clear of the pouch, curl the ring finger and little finger underneath the magazine and rotate it up to observe rounds in the magazine.

Load-bearing Vest

With the right hand, withdraw magazines from the left side of the vest (see fig. 3-11). With the left hand, withdraw magazines from the right side of the vest (see fig. 3-12).

To withdraw magazines from a load-bearing vest:

- With the thumb and index finger, unfasten the snap on the vest pouch.
- Slide the thumb over the magazine, feeling for a base plate indicating a filled magazine.
- Rotate the hand over the magazine while sliding the thumb to the back of the magazine.
- Grasp the magazine with the thumb, index finger, and middle finger, while curling the ring finger and little finger on top of the base plate.

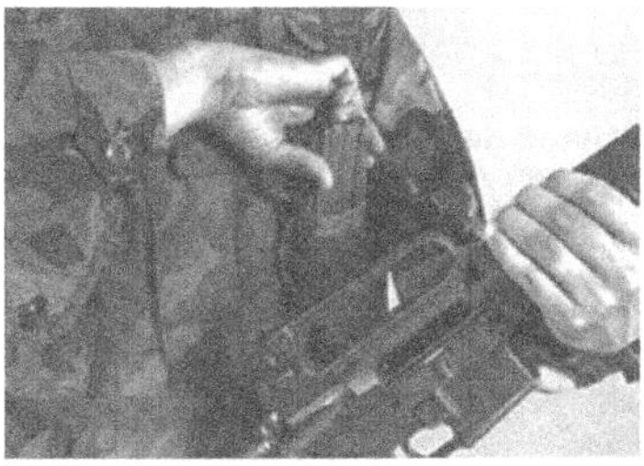

Figure 3-11. Withdrawing a Magazine with the Right Hand.

- Lift the magazine directly out of the pouch and rotate it up to observe the rounds in the magazine.

3006. Reloading the Rifle

Principles of Reloading

The first priority when performing a reload is to get the rifle reloaded and back into action. The second priority when performing a reload is to retain the magazine so when you move, the magazine moves with you. When time permits, retain magazines securely on your person (e.g., in magazine pouch, flak jacket, and cargo pocket). The combat situation may dictate dropping the magazine to the deck when performing a reload. This is acceptable as long as it is picked up before moving to another location.

Take cover before reloading. Always reload before leaving cover to take advantage of the protection provided by cover.

When moving, your focus should be on moving, therefore every effort should be made to not reload while on the move.

When reloading, your focus is on the magazine change.

When reloading, draw the weapon in close to your body so you can see what you are doing and retain positive control of the magazine.

When the new magazine is inserted, tug on it to ensure it is seated. Do not slam the magazine into the weapon hard enough to cause a round to partially pop out of the magazine. This action will cause a double feed and require remedial action.

Figure 3-12. Withdrawing a Magazine

Retain your empty magazines. When there is a lull in the action, refill those magazines so they will be available for future use.

During a lull in the action, replace your magazine when you know you are low on ammunition. This ensures a full magazine of ammunition in the rifle should action resume. Do not wait until the magazine is completely empty to replace it.

Condition 1 Reload

A condition 1 reload is performed when the weapon is in condition 1 by replacing the magazine before it runs out of ammunition. To perform a condition 1 reload, perform the following steps:

- Withdraw a filled magazine from the magazine pouch. With the same hand, press the magazine button and remove the partially filled magazine so it can be retained in the remaining fingers.
- Fully insert the filled magazine into the magazine well and tug downward on the magazine to ensure it is properly seated.
- Store the partially filled magazine in the magazine pouch with rounds up and projectiles pointing away from the body.
- Fasten the magazine pouch.

Dry Reload

A dry reload is required when the magazine in the weapon has been emptied and the bolt has locked to the rear. To perform a dry reload—

- Press the magazine release button.
- Remove the empty magazine and retain it on your person when time permits.
- Fully insert a filled magazine into the magazine well and tug downward on the magazine to ensure it is properly seated.
- Depress the bolt catch to allow the bolt carrier to move forward and observe the round being chambered. This places the rifle in Condition 1.

3007. Remedial Action

If the rifle fails to fire, a Marine performs remedial action. Remedial action is the process of investigating the cause of the stoppage, clearing the stoppage, an returning the weapon to operation.

Observe for Indicators

Once the rifle ceases firing, the Marine must visuall or physically observe the ejection port to identify th problem before he can clear it. The steps taken to clea the weapon are based on observation of one of th following three indicators:

Indicator: The bolt is forward or the ejection por cover is closed. See figures 3-13 and 3-14.

Figure 3-13. Bolt Forward.

Figure 3-14. Ejection Port Cover Closed.

To return the weapon to operation—

- Seek cover if the tactical situation permits.
- Tap—Tap the bottom of the magazine.
- Rack—Pull the charging handle to the rear an release it.
- Bang—Sight in and attempt to fire.

Indicator: Brass is obstructing chamber area (usually indicating a double feed or failure to eject). See figure 3-15.

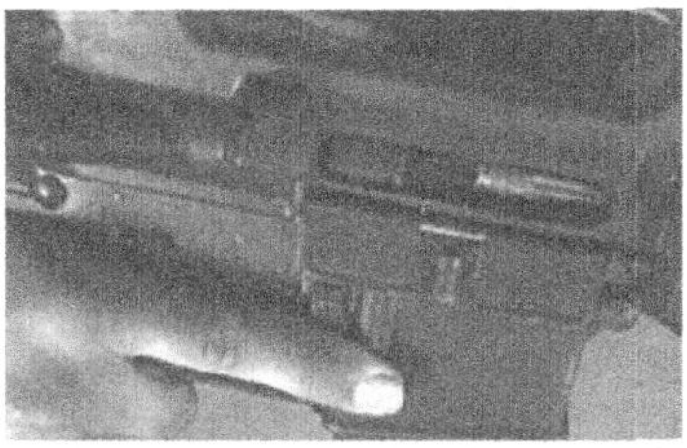

Figure 3-15. Brass Obstructing the Chamber.

To return the weapon to operation—

- Seek cover if the tactical situation permits.
- Attempt to remove the magazine.
- Attempt to lock the bolt to the rear

If the bolt will not lock to the rear, rotate the rifle so the ejection port is facing down; hold the charging handle to the rear as far as it will go and shake the rifle to free the round(s). If the rounds do not shake free, hold the charging handle to the rear and strike the butt of the rifle on the ground or manually clear the round. Conduct a reload. Sight in and attempt to fire.

Indicator: The bolt is locked to the rear. See figure 3-16. To clear return the weapon to operation—

Figure 3-16. Bolt Locked to the Rear.

Note

Although a dry weapon is not considered a true stoppage or mechanical failure, the Marine must take action to return the weapon to operation.

- Seek cover if the tactical situation permits.
- Conduct a dry reload.
- Sight in and attempt to fire.

Audible Pop or Reduced Recoil

An audible pop occurs when only a portion of the propellant is ignited. It is normally identifiable by reduced recoil and is sometimes accompanied by excessive smoke escaping from the chamber area. To clear the rifle in a combat environment:

- Place the rifle in Condition 4.
- Move take down pin from left to right as far as it will go to allow the lower receiver to pivot.
- Remove the bolt carrier group.
- Inspect the bore for an obstruction from the chamber end.
- Insert a cleaning rod into the bore from the muzzle end and clear the obstruction.
- Reassemble the rifle.
- Conduct a reload.
- Sight in and attempt to fire.

3008. Weapons Carries

Weapons carries provide an effective way to handle the rifle while remaining alert to enemy engagement. Weapons carries are tied to threat conditions and are assumed in response to a specific threat situation. The weapons carry assumed prepares the Marine, both mentally and physically, for target engagement. The sling provides additional support for the weapon when firing; therefore, the hasty sling should be used in conjunction with the carries.

Tactical Carry

A Marine carries the rifle at the tactical carry if no immediate threat is present. The tactical carry permits control of the rifle while a Marine is moving, yet it still allows quick engagement of the enemy. A Marine performs the following steps to assume the tactical carry:

- Place left hand on the handguards, right hand around the pistol grip, trigger finger straight along the receiver (see fig. 3-17), and right thumb on top of the selector lever (see fig. 3-18).

Figure 3-17. Straight Trigger Finger.

Figure 3-18. Thumb on Selector Lever.

- Place the buttstock along the side of the body at approximately hip level.
- Angle the muzzle upward about 45 degrees in a safe direction.
- Position the muzzle in front of the eyes, slightly below eye level (see fig. 3-19).
- Move the head and the eyes with the muzzle as it moves.

Figure 3-19. Tactical Carry.

Alert Carry

A Marine carries the rifle at the alert if enemy contact is likely. The alert is also used for moving in close terrain (e.g., urban, jungle). A Marine can engage the enemy faster from the alert than from the tactical carry. However, the alert is more tiring than the tactical carry and its use can be physically demanding. A Marine performs the following steps to assume the alert:

- Place the left hand on the handguards, the right hand around the pistol grip, the trigger finger straight along the receiver (see fig. 3-17), and the right thumb on top of the selector lever (see fig. 3-18).
- Place the buttstock in the shoulder.
- Angle the muzzle downward about 45 degrees and point it in a safe direction or the general direction of likely enemy contact (see fig. 3-20).

Figure 3-20. Alert Carry.

Ready Carry

A Marine carries the rifle at the ready if contact with the enemy is imminent. The ready allows immediate target engagement, but it is very tiring to maintain over a long period of time. A Marine performs the following steps to assume the ready:

- Place left hand on handguards, right hand around the pistol grip, the trigger finger straight along the receiver (see fig. 3-17), and the right thumb on top of the selector lever (see fig. 3-18).
- Place the buttstock in the shoulder.
- Point the muzzle in the direction of the enemy.

- Lower the sights to just below eye level so that a clear field of view is maintained so that a target may be identified (see fig. 3-21).

Figure 3-21. Ready Carry.

3009. Weapons Transports

Weapons transports are used to carry the rifle over the back or shoulders when moving for long periods; they provide a more relaxed position for walking. Weapons transports are used if no immediate threat is present. They are also used whenever one or both hands are needed for other work.

Strong Side Sling Arms Transport (Muzzle Up)

To assume the strong side sling arms (muzzle up) transport from the tactical carry, a Marine performs the following steps (see fig. 3-22):

- Release the hold on the pistol grip.
- Lower the buttstock and bring the rifle to a vertical position.
- With the right hand, grasp the sling above the left forearm.

Figure 3-22. Strong Side Sling Arms Transport (Muzzle Up).

- With the left hand, guide the rifle around the right shoulder.
- With the right hand, apply downward pressure on the sling. This stabilizes the rifle on the shoulder.

Weak Side Sling Arms Transport (Muzzle Down)

The weak side sling arms (muzzle down) transport can be used in inclement weather to keep moisture out of the rifle's bore. To assume this transport from the tactical carry, a Marine performs the following steps (see fig. 3-23):

- Release the hold on the pistol grip.
- With the left hand, rotate muzzle down and bring the rifle to a vertical position on the left side of the body. The pistol grip is pointed outboard.
- With right hand, place sling on left shoulder.
- Grasp sling above the waist with the left hand.
- With the left hand, apply downward pressure on the sling. This stabilizes the rifle on the shoulder.

Figure 3-23. Weak Side Sling Arms Transport (Muzzle Down.)

Cross Body Sling Arms Transport

A Marine uses the cross body sling arms transport if he requires both hands for work. The rifle is slung across the back with the muzzle up or down. Normally, the rifle is carried with the muzzle down to prevent pointing the muzzle in an unsafe direction.

To assume a cross body sling transport, a Marine performs the following steps from weak side sling arms (muzzle down) (see fig. 3-24):

Figure 3-24. Cross Body Sling Arms Transport (Muzzle Down).

- With the right hand, grasp the sling.
- With the left hand, grasp the handguards.
- Pull up on the rifle with both hands.
- Slide the sling over the head.
- Position the rifle so that it rests comfortably across the back.

To assume a cross body sling transport, a Marine performs the following steps from strong side sling arms (muzzle up) (see fig. 3-25):

- With the left hand, grasp the sling.
- With the right hand, grasp the pistol grip.
- Pull up on the rifle with both hands.
- Slide sling over head.
- Position the rifle so that it rests comfortably across the back.

Figure 3-25. Cross Body Sling Arms. Transport (Muzzle Up).

Note

Ensure the muzzle of the rifle is maintained in a safe direction when assuming this transport.

3010. Transferring the Rifle

Proper weapons handling is required every time a Marine picks up a weapon, passes a weapon to another Marine, or receives a weapon from another Marine. It is the responsibility of the Marine receiving or taking charge of a weapon to determine its condition. Depending on the situation, there are two procedures that can be used to transfer a rifle from one Marine to another Marine: show clear transfer and condition unknown transfer.

Show Clear Transfer

When time and the tactical situation permit, the Marine should transfer the rifle using the show clear transfer. To properly pass a rifle between Marines, the Marine handing off the rifle must perform the following procedures:

- Ensure the rifle is on safe.
- Remove the magazine if it is present.
- Lock the bolt to the rear.
- Visually inspect the chamber to ensure there is no ammunition present.
- Leave the bolt locked to the rear and hand the weapon to the other Marine.

The Marine receiving the weapon must:

- Ensure the rifle is on safe.
- Visually inspect the chamber to ensure there is no ammunition present.
- Release the bolt catch and observe the bolt going forward on an empty chamber.
- Close the ejection port cover.

Condition Unknown Transfer

There are times when time or the tactical situation does not permit a show clear transfer of the rifle. The procedures for the condition unknown transfer are conducted by a Marine when he takes charge of a rifle in any situation when the condition of the rifle is unknown (e.g., an unmanned rifle from a casualty, a rifle stored in a rifle rack). To properly take charge of a rifle when its condition is unknown, the Marine must perform the following procedures:

- Ensure the rifle is on safe.
- Conduct a chamber check to determine the condition of the weapon (see para. 3003).
- Remove the magazine and observe if ammunition is present. If time permits, count the rounds.
- Insert the magazine into the magazine well.
- Close the ejection port cover.

3011. Clearing Barrel Procedures

Purpose of a Clearing Barrel

The sole purpose of a clearing barrel is to provide a safe direction in which to point a weapon when loading; unloading; and unloading and showing clear. See figure 3-26 on page 3-12. Clearing barrel procedures are identical to the weapons handling procedures for the rifle for the loading; unloading; and unloading and showing clear.

Figure 3-26. Clearing Barrel.

Procedures for "Load"

On the command "Load," the Marine will perform the following steps to take the rifle from Condition 4 to Condition 3:

- With a straight trigger finger, point the rifle in the clearing barrel.
- Ensure the rifle is on safe.
- Withdraw a magazine from the magazine pouch.
- Observe the magazine to ensure it is filled.
- Fully insert the magazine into the magazine well.
- Without releasing the magazine, tug downward on the magazine to ensure it is seated.
- Fasten the magazine pouch.
- Close the ejection port cover.

Procedures for "Make Ready"

If standard operating procedures (SOP) or rules of engagement (ROE) require the rifle to be carried in Condition 1, the Marine will "Make Ready" at the clearing barrel. On the command "Make Ready," the Marine will perform the following steps to take the rifle from Condition 3 to Condition 1:

- Pull the charging handle to the rear and release. There are two methods of doing this:
  - Grip the pistol grip firmly with the right hand and pull the charging handle with the left hand to its rearmost position and release.
  - Or grip the handguards firmly with the left hand and pull the charging handle with the right hand to its rearmost position and release.
- To ensure ammunition has been chambered, conduct a chamber check.
- Check the sights (to ensure proper BZO setting, correct rear sight aperture, etc.).
- Close the ejection port cover.

Procedures for "Unload"

On the command "Unload," the Marine will perform the following steps to take the rifle from any condition to Condition 4:

- With a straight trigger finger, point the rifle in the clearing barrel.
- Ensure the weapon is on safe.
- Remove the magazine from the rifle and retain it on your person.
- While cupping the left hand under the ejection port, rotate weapon until the ejection port is facing down.
- Pull the charging handle to the rear and catch the round in the left hand (see fig. 3-17).
- Lock the bolt to the rear.
- Put weapon on safe if not already on safe.
- Ensure that the chamber is empty and that no ammunition is present.
- Release the charging handle and observe the bolt moving forward on an empty chamber.
- Close the ejection port cover.
- Check the sights (for proper BZO setting, correct rear sight aperture, etc.).
- Place any ejected round into the magazine and return the magazine to the magazine pouch and close the magazine pouch.

Procedures for "Unload and Show Clear"

On the command "Unload and Show Clear," the Marine will perform the following steps to take the rifle from any condition to Condition 4:

- With a straight trigger finger, point the rifle in the clearing barrel.
- Ensure the weapon is on safe.
- Remove the magazine from the rifle and retain it on your person.
- While cupping the left hand under the ejection port, rotate the weapon until the ejection port is facing down.
- Pull the charging handle to the rear and catch the round in the left hand.
- Lock the bolt to the rear.

- Ensure the chamber is empty and no ammunition is present.
- Have an observer inspect the weapon to ensure no ammunition is present. The observer:
  - Visually inspects chamber to ensure it is empty, no ammunition is present, and the magazine is removed.
  - Ensures the weapon is on safe.
  - Acknowledges the rifle is clear.
- After receiving acknowledgment from the observe that the rifle is clear, the Marine releases the bol catch and observes the bolt moving forward on a empty chamber.
- Close the ejection port cover.
- Check the sights (for proper BZO setting, correc rear sight aperture, etc.).
- Place any ejected round into the magazine an return the magazine to the magazine pouch an close the magazine pouch.

CHAPTER 4. FUNDAMENTALS OF MARKSMANSHIP

The fundamentals of marksmanship are aiming, breathing, and trigger control. These techniques provide the foundation for all marksmanship principles and skills. For rifle fire to be effective, it must be accurate. A rifleman who merely sprays shots in the vicinity of the enemy produces little effect. The fundamentals of marksmanship, when applied correctly, form the basis for delivering accurate fire on enemy targets. These skills must be developed so that they are applied instinctively. During combat, the fundamentals of marksmanship must be applied in a time frame consistent with the size and the distance of the target. At longer ranges, the target appears to be smaller and a more precise shot is required to accurately engage the target. The fundamentals are more critical to accurate engagement as the range to the target increases. To be accurate at longer ranges, the Marine must take the time to slow down and accurately apply the fundamentals. At shorter ranges, the enemy must be engaged quickly before he engages the Marine. As the size of the target increases, and the distance to the target decreases, the fundamentals, while still necessary, become less critical to accuracy.

Note

☞The procedures in this manual are written for right-handed Marines; left-handed Marines should reverse instructions as necessary.

4001. Aiming

Sight Alignment

Sight alignment is the relationship between the front sight post and rear sight aperture and the aiming eye. This relationship is the most critical to aiming and must remain consistent from shot to shot. To achieve correct sight alignment (see fig. 4-1):

- Center the tip of the front sight post vertically and horizontally in the rear sight aperture.
- Imagine a horizontal line drawn through the center of the rear sight aperture. The top of the front sight post will appear to touch this line.
- Imagine a vertical line drawn through the center of the rear sight aperture. The line will appear to bisect the front sight post.

Sight Picture

Sight picture is the placement of the tip of the front sight post in relation to the target while maintaining sight alignment. Correct sight alignment but improper sight placement on the target will cause the bullet to impact the target incorrectly on the spot where the sights were aimed when the bullet exited the muzzle.

To achieve correct sight picture, place the tip of the front sight post at the center of the target while maintaining sight alignment (see fig. 4-2). Center mass is

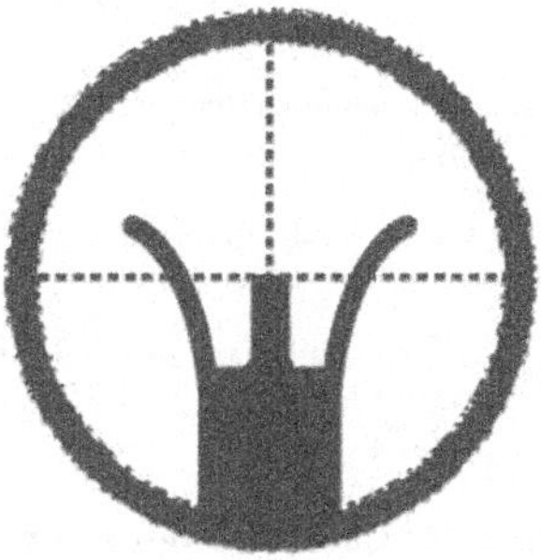

Figure 4-1. Correct Sight Alignment.

Figure 4-2. Correct Sight Picture.

Figure 4-3. Examples of Correct Sight Picture.

the correct aiming point so that point of aim/point of impact is achieved.

The sighting system for the M16A2 rifle is designed to work using a center mass sight picture.

In combat, targets are often indistinct and oddly shaped. The center mass hold provides a consistent aiming point (see fig. 4-3).

Importance of Correct Sight Alignment

A sight alignment error results in a misplaced shot. The error grows proportionately greater as the distance to the target increases. An error in sight picture, however, remains constant regardless of the distance to the target. See figure 4-4.

Factors Affecting Sight Alignment and Sight Picture

Stock Weld

Stock weld is the point of firm contact between the cheek and the stock of the rifle. See figure 4-5. The head should be as erect as possible to enable the aiming eye to look straight through the rear sight aperture. If the position of the Marine's head causes him to look across the bridge of his nose or out from under his eyebrow, the eye will be strained. The eye functions best in its natural forward position. Changing the placement of the cheek up or down on the stock from shot to shot may affect the zero on the rifle due to the perception of the rear sight aperture. A consistent and proper stock weld is critical to the aiming process because it provides consistency in eye relief, which affects the ability to align the sights.

Eye Relief

Eye relief is the distance between the rear sight aperture and the aiming eye. See figure 4-6. Normal eye relief is two to six inches from the rear sight aperture. The distance between the aiming eye and the rear sight aperture depends on the size of the Marine and the firing position. While eye relief varies slightly from one position to another, it is important to have the same eye relief for all shots fired from a particular position.

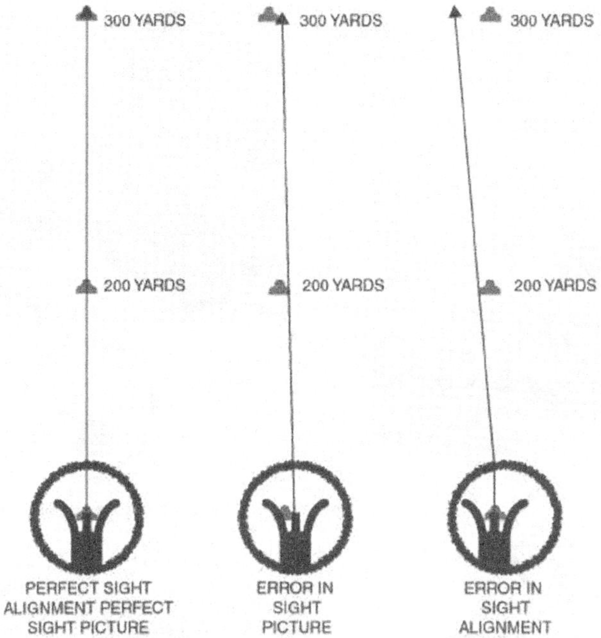

Figure 4-4. Importance of Correct Sight Alignment.

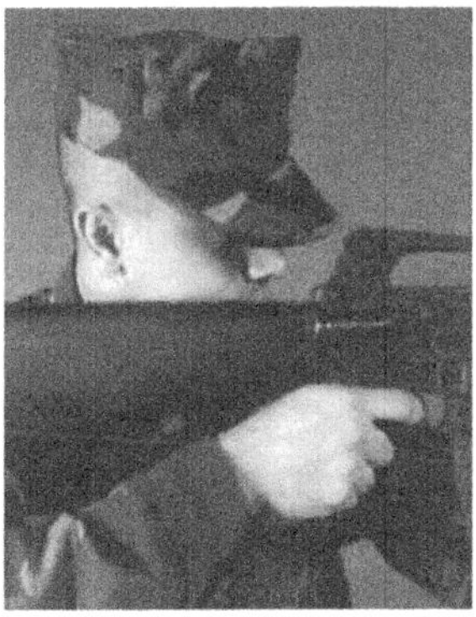

Figure 4-5. Stock Weld.

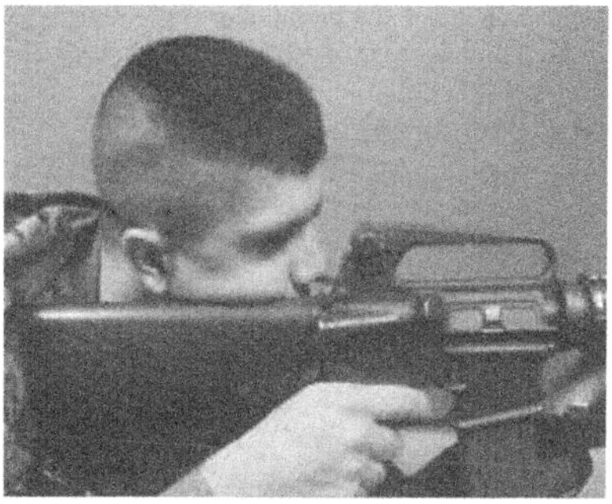

Figure 4-6. Proper Eye Relief.

If the eye is too close to the rear sight aperture, it will be difficult to line up the front sight post in the rear sight aperture (see fig. 4-7). Moving the eye back from the rear sight aperture will make the aperture appear smaller and allow the tip of the front sight post to be easily lined up inside the rear sight aperture.

Figure 4-7. Shortened Eye Relief.

If the eye is too far from the rear sight aperture, it will be difficult to acquire the target and to maintain a precise aiming point (see fig. 4-8).

Wearing of Glasses

Wearing glasses can alter the perception of sight alignment and sight picture. If wearing glasses, it is critical to look through the optic center of the lens.

Acquiring and Maintaining Sight Alignment and Sight Picture

The human eye can focus clearly on only one object at a time. For accurate shooting, it is important to focus on the tip of the front sight post. When the shot is fired, focus must be on the tip of the front sight post; peripheral vision will include the rear sight and the target. The rear sight and the target will appear blurry. Staring or fixing the vision on the front sight post for longer than a few seconds can distort the image, making it difficult to detect minute errors in sight alignment.

Figure 4-8. Extended Eye Relief.

Proper stock weld and placement of the rifle butt in the shoulder aid in establishing sight alignment quickly. The rifle butt's placement in the shoulder serves as the pivot point for presenting the rifle up to a fixed point on the cheek (stock weld). During combat, a Marine will look at the target as the rifle is presented. As rifle sights become level with the aiming eye, a Marine visually locates the target through the rear sight aperture. As the rifle settles, a Marine's focus shifts back to the front sight post to place the tip of the post on the target and obtain sight alignment and sight picture. To maintain sight alignment and sight picture, the Marine's focus should shift repeatedly from the front sight post to the target until correct sight alignment and sight picture are obtained. This enables the detection of minute errors in sight alignment and sight picture.

Size and Distance to the Target

During combat, the fundamentals of marksmanship must be applied in a time frame consistent with the size and the distance to the target.

Long-range Engagements

At longer ranges, the target appears smaller and a more precise shot is required to accurately engage the target. Sight alignment and sight picture are more critical to accurate engagement as the range to the target increases. To be accurate at longer ranges, the Marine must take the time to slow down and accurately apply the fundamentals.

As the distance to the target increases, the front sight post covers more of the target, making it difficult to establish a center of mass hold (see fig. 4-9). Since the Marine must see the target to engage it, there is a tendency to look at the target by lowering the tip of the front sight post. This causes shots to impact low or miss the target completely. A Marine must consciously aim at the center of mass and attempt to maintain a center mass sight picture.

Short-range Engagements

At shorter ranges, the enemy must be engaged quickly before he engages the Marine. As distance to the target decreases, the size of the target appears to increase, and sight alignment becomes less critical to accuracy. At very short ranges, a deviation in sight alignment can still produce accurate results as long as the tip of the front sight post is in the rear sight aperture and on the target (see fig. 4-10). The time required to engage a target is unique to each individual. Although a Marine must engage the target rapidly, some semblance of sight alignment is still required to be accurate.

4002. Breath Control

Proper breath control is critical to the aiming process. Breathing causes the body to move. This movement transfers to the rifle making it impossible to maintain proper sight picture. Breath control allows the Marine to fire the rifle at the moment of least movement.

Breath Control During Long-range or Precision Fire (Slow Fire)

It is critical that Marines interrupt their breathing at a point of natural respiratory pause before firing a long-range shot or a precision shot from any distance. A respiratory cycle lasts 4 to 5 seconds. Inhaling and exhaling each require about 2 seconds. A natural pause of 2 to 3 seconds occurs between each respiratory cycle. The pause can be extended up to 10 seconds. During the pause, breathing muscles are relaxed and the sights settle at their natural point of aim. To minimize movement, Marines must fire the shot during the natural respiratory pause. The basic technique is as follows:

- Breathe naturally until the sight picture begins to settle.
- Take a slightly deeper breath.
- Exhale and stop at the natural respiratory pause.
- Fire the shot during the natural respiratory pause.

Note

If the sight picture does not sufficiently settle to allow the shot to be fired, resume normal breathing and repeat the process.

Breath Control During All Other Combat Situations

A Marine in a combat environment may not have the time to fire a shot during the natural respiratory pause.

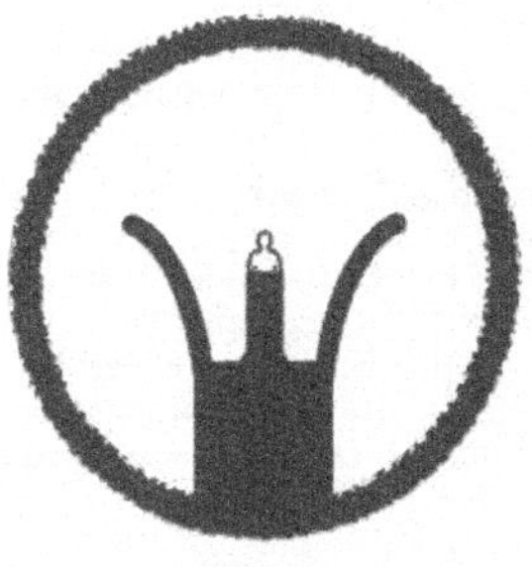

Figure 4-9. Sight Picture at Long-range Engagements.

Figure 4-10. Sight Picture at Short-range Engagements.

It may be necessary to take several deep breaths before holding the breath. A Marine should not make an exaggerated effort to perform breath control. A natural respiratory pause will help stabilize the shooter's sight picture. The basic technique is as follows:

- Take a deep breath filling the lungs with oxygen.

Note

It may be necessary to take several deep breaths quickly before holding the breath.

- Hold the breath and apply pressure to the trigger.
- Fire the shots.

4003. Trigger Control

Trigger control is the skillful manipulation of the trigger that causes the rifle to fire without disturbing sight alignment or sight picture. Controlling the trigger is a mental process, while pulling the trigger is a physical process.

Grip

A firm grip is essential for effective trigger control. The grip is established before starting the application of trigger control and it is maintained through the duration of the shot. To establish a firm grip on the rifle, position the "V" formed between the thumb and index finger on the pistol grip behind the trigger. The fingers and the thumb are placed around the pistol grip in a location that allows the trigger finger to be placed naturally on the trigger and the thumb in a position to operate the safety. Once established, the grip should be firm enough to allow manipulation of the trigger straight to the rear without disturbing the sights. See figure 4-11.

Trigger Finger Placement

Correct trigger finger placement allows the trigger to be pulled straight to the rear without disturbing sight alignment. The trigger finger should contact the trigger naturally. The trigger finger should not contact the rifle receiver or trigger guard.

Figure 4-11. Grip.

Types of Trigger Control

There are two techniques of trigger control: uninterrupted and interrupted

Uninterrupted Trigger Control

The preferred method of trigger control in a combat environment is uninterrupted trigger control. After obtaining sight picture, the Marine applies smooth, continuous pressure rearward on the trigger until the shot is fired.

Interrupted Trigger Control

Interrupted trigger control is used at any time the sight alignment is interrupted or the target is temporarily obscured. An example of this is extremely windy conditions when the weapon will not settle, forcing the Marine to pause until the sights return to his aiming point. To perform interrupted trigger control:

- Move the trigger to the rear until an error is detected in the aiming process.
- When this occurs, stop the rearward motion on the trigger, but maintain the pressure on the trigger, until sight picture is achieved.
- When the sight picture settles, continue the rearward motion on the trigger until the shot is fired.

Resetting the Trigger

During recovery, release the pressure on the trigger slightly to reset the trigger after the first shot is delivered (indicated by an audible click). Do not remove the finger from the trigger. This places the trigger in position to fire the next shot without having to reestablish trigger finger placement.

4004. Follow-Through/Recovery

Follow-Through

Follow-through is the continued application of the fundamentals until the round has exited the barrel. In combat, follow-through is important to avoid altering the impact of the round by keeping the rifle as still as possible until the round exits the barrel.

Recovery

It is important to get the rifle sights back on the target for another shot. This is known as recovery. Shot recovery starts immediately after the round leaves the barrel. To recover quickly, a Marine must physically bring the sights back on target as quickly as possible.

Chapter 5. Rifle Firing Positions

In a combat environment, a Marine must be prepared to engage the enemy under any circumstance. There are four basic firing positions: prone, sitting, kneeling, and standing. These positions provide a stable foundation for effective shooting. Any firing position must provide stability, mobility, and observation of the enemy. During training, a Marine learns positions in a step-by-step process, guided by a series of precise movements until the Marine assumes a correct position. The purpose of this is to ensure that the Marine correctly applies all of the factors that assist him in holding the rifle steady. The Marine will gradually become accustomed to the feel of the positions through practice and eventually will be able to know instinctively whether his position is correct. In combat, it may not be possible to assume a textbook firing position due to terrain, available cover, engagement time, dispersion of targets, and other limiting factors. Modifications to the basic positions may have to be made to adjust to the combat environment. The Marine must strive to assume a position that offers stability for firing, maximum cover and concealment from the enemy, and maximum observation of the target.

Note

+The procedures in this manual are written for right-handed Marines; left-handed Marines should reverse instructions as necessary.

5001. Selecting a Firing Position

The selection of a firing position (prone, sitting, kneeling, standing) is based on terrain, available cover, dispersion of targets, and other limiting factors. A Marine must select a position that offers stability, mobility, and observation.

Stability

A firing position must provide a stable platform for accurate and consistent shooting. If the position is solid, the front sight can be held steady and the rifle sights should recover after recoil to the same position on the target. This allows for rapid reengagement of the enemy. The prone position provides the most stability for firing, while the standing position provides the least stability.

Mobility

A firing position must provide a Marine with the mobility required to move to new cover or to another area. The standing position permits maximum mobility, and it also allows the most lateral movement for engagement of widely dispersed targets. The prone position allows the least mobility and allows limited lateral movement.

Observation of the Enemy

A firing position must limit a Marine's exposure to the enemy, yet allow observation of the enemy. Manmade structures and terrain features (e.g., vegetation, earth contours) often dictate the shooting position. The standing position normally provides the best field of view, but it usually allows the most exposure to the enemy. The prone position normally allows the least exposure, but it usually provides a limited field of view.

5002. Types and Uses of the Rifle Web Sling

The rifle sling, when adjusted properly, provides maximum stability for the weapon, and helps hold the front sight still and reduce the effects of the rifle's recoil. Once a sling adjustment is found that provides maximum control of the weapon, the same sling adjustment should be maintained. Varying the sling tension extensively will affect the strike of the bullet, which will make maintaining a BZO difficult. Using the same sling adjustment will ensure the accuracy of rounds on target. There are two basic types of rifle sling adjustments: the hasty sling and loop sling.

Note

In training situations, the parade sling may be used. See paragraph 5008.

Hasty Sling

Application

The hasty sling is used in all firing positions. The hasty sling is advantageous in combat because it can be acquired quickly and it provides added stability to the rifle. The same sling setting can be used for all firing positions. If properly adjusted, the hasty sling supports the weight of the weapon, provides stability for the rifle, and reduces the effects of the rifle's recoil. When using the hasty sling, controlled muscle tension is applied to offer resistance against the sling, enabling the rifle sights to be held steady.

Donning the Hasty Sling

A Marine performs the following steps to form a hasty sling:

Figure 5-1. J-hook Location.

- Hold the rifle vertical with the barrel pointing upward.
- Unhook the J-hook from the lower sling swivel.
- Loosen the sling keeper.
- Adjust the sling until the J-hook hangs below the rifle butt. (The distance will vary based on the individual Marine, but the J-hook will usually hang approximately 3 to 10 inches below the rifle butt.) Secure the sling keeper. See figure 5-1.
- Turn the sling a half turn outboard to allow the sling to lay flat against the arm.
- Attach the J-hook to the lower sling swivel so the open end of the J-hook faces outboard, away from the rifle. See figure 5-2.

Figure 5-2. J-hook Turned Outboard.

- While holding the rifle with the right hand, place the left arm through the sling near the lower sling swivel. Slide the arm up through the sling below the half twist. The sling makes contact low on the arm just below the triceps, above the elbow. The sling lies flat on the back of the arm.
- With the left hand, grasp the handguard by pinching it in the "V" formed by the thumb and forefinger. The sling lies flat against the back or side of the wrist or on the arm near the wrist. See figure 5-3.

Figure 5-3. Sling Against the Wrist.

- Move the left hand as required to level the rifle with the line of sight. Placement of the forward hand controls the tension on the sling between the back of the wrist or arm and the upper sling swivel (see fig. 5-4). This hand placement, with a straight locked wrist, will cause the sling to pull straight under the handguards and serves to stabilize the front sight of the rifle.

- Move the feed end of the sling in or out of the sling keeper to adjust the hasty sling. Sling tension is in-

Figure 5-4. Position of Forward Hand.

creased by pushing the elbow outboard. See figure 5-5. (This enables one sling setting to fit all positions.) It is important for the hasty sling to be adjusted so it supports the rifle. The sling setting must allow the left elbow to push outboard against the sling so the elbow is not inverted under the rifle.

Figure 5-5. Position of Left Elbow.

- Locate the sling keeper near the feed end of the sling and secure so the backside or flat end of the sling keeper is against the arm.

Loop Sling

Application

The loop sling provides the greatest amount of stability during firing. This stability allows the Marine to perfect marksmanship fundamentals. A loop sling takes longer to don or remove than a hasty sling. Therefore, it has limited combat application and is best used where stability of hold is needed for a precision or long-range shot. The loop sling is used in the prone, sitting, and kneeling positions.

Donning the Loop Sling

- Place the rifle butt on the right hip and cradle the rifle in the right arm.
- Disconnect the J-hook from the lower sling swivel.
- With the M-buckle near the hook, feed the sling through the top of the M-buckle to form a loop large enough to slip over the arm. See figure 5-6.
- Give the loop a half turn outboard and insert the left arm through the loop, positioning the loop above the biceps. The loop is high on the left arm above the biceps muscle in such a position that it does not transmit pulse beat to the rifle.
- Position the M-buckle on the outside of the left arm. See figure 5-7.
- Tighten the loop on the left arm, ensuring the M-buckle moves toward the center of the arm as the loop tightens. The sling must pull from the center of the arm to be properly positioned. In this way, as tension is applied to the sling in the firing position, the loop will tighten.

To adjust the sling for the proper length, loosen the sling keeper and pull up or down (toward or away) from the loop. This adjustment varies with every individual and every firing position. The loop should not be tightened excessively on the arm. If blood flow is restricted, excessive pulse beat is transmitted through the rifle sling to the rifle and causes a noticeable, rhythmic movement of the rifle sights. When this occurs, a stable hold at the desired aiming point is impossible to achieve.

Figure 5-6. Forming a Loop.

Figure 5-7. Position of M-buckle.

Tension on the rifle sling is correct when it causes the rifle butt to be forced rearward into the pocket of the shoulder. This serves to keep the butt plate in the shoulder pocket during recoil. To increase tension on the rifle sling, the sling must be shortened. To lessen the tension, the rifle sling must be lengthened.

Move the sling keeper toward the left arm and secure it. The sling keeper should be positioned near the feed end of the sling.

Place the left hand over the sling from the left side and under the rifle. The rifle handguard should rest in the "V" formed between the thumb and forefinger and across the palm of the hand.

Move the left hand as required to achieve desired sight picture. Adjust the length of the sling for proper sling tension and support. See figure 5-8.

Figure 5-8. Loop Sling Donned.

5003. Factors Common to All Shooting Positions

There are seven factors that are common to all shooting positions that affect the ability to hold the rifle steady, maintain sight alignment, and control the trigger. The way these factors are applied differs slightly for each position, but the principles of each factor remain the same.

Left Hand

Placement of the left hand affects placement of left elbow, eye relief, stock weld, and sling tension.

Hasty Sling

In a hasty sling configuration, the sling is attached to the upper and lower sling swivels of the rifle. When the left arm is placed in the hasty sling, tension created by the sling travels from side to side. The tension created by the sling affects how the position is established. There are fundamental differences between the application of the seven factors when using the hasty sling. The most obvious of these is placement of the left hand and the left elbow.

To maximize the support provided by the hasty sling, the left elbow should not be inverted and under the rifle. Instead, the elbow should be pushed outboard against the sling. To achieve this, the position of the shooter's body must almost face the target as opposed to being perpendicular to the target. In addition, the hasty sling must be loosened to allow the elbow to push out against the sling far enough so that the elbow is not under the rifle.

The tension on the sling created by the hasty sling causes the center of balance to change on the rifle. When the elbow is under the rifle with the hasty sling donned, the sling pulls down on the sling swivel disrupting the center of balance and causing the muzzle to drop. Therefore, the elbow must be pushed outboard.

Outboard tension on the sling by the elbow drives the buttstock into the pocket of the shoulder. To enable this, the sling must make contact on the arm just below the triceps, above the elbow. See figure 5-9.

Figure 5-9. Position of Left Elbow with Hasty Sling.

To stabilize the front sight of the rifle, the forward hand, wrist, and forearm should be straight with the wrist locked in place; the hand should be rotated up so the rifle rests in the "V" formed by the thumb and index finger; the fingers will not curl around the handguards. Instead, they will pinch the handguard slightly to keep the hand from slipping on the handguard during recoil. See figure 5-10.

Figure 5-10. Position of Left Hand with Hasty Sling.

When the wrist of the left hand is straight and locked, it creates resistance on the sling close to the muzzle. The sling is in contact with the back or side of the wrist or on the arm near the wrist. This resistance allows the front sight to be stabilized.

In contrast, when the rifle rests across the palm of the hand, the only resistance created is where the sling meets the triceps. Since the resistance is further from the muzzle of the rifle, it makes stabilizing the front sight more difficult.

Loop Sling

With the loop sling donned, the handguard of the rifle rests in the "V" formed by the thumb and index finger of the left hand. The rifle rests across the heel of the hand. The left elbow should be positioned under the weapon to create bone support and a consistent resistance to recoil. The fingers can curl around the handguard, but should apply only the minimum amount of pressure to prevent the hand from slipping on the handguard. See figure 5-11.

Figure 5-11. Position of Left Hand with Loop Sling.

Rifle Butt in the Pocket of the Shoulder

The rifle butt placed firmly in the pocket formed in th right shoulder provides resistance to recoil, help steady the rifle, and prevents the rifle butt from slip ping in the shoulder during firing. Consistent place ment of the rifle butt in the shoulder pocket is essentia to maintaining a BZO and firing tight shot groups.

Hasty Sling

With the hasty sling donned, the butt of the rifle i placed in the pocket of the shoulder. The body i squared to the target to provide a pocket for the butt o the weapon.

Outboard tension on the sling by the left elbow drive the buttstock into the pocket of the shoulder.

The rifle butt should be placed high in the shoulder t achieve a proper stock weld, allowing the Marine t bring the stock up to his head, rather than lower hi head to the stock, which can degrade acquisition o sight alignment and sight picture. See figure 5-12.

Figure 5-12. Buttstock in the Shoulder with Hasty Sling.

Loop Sling

With the loop sling donned, the toe of the rifle butt i placed in the pocket of the shoulder. See figure 5-13.

Figure 5-13. Buttstock in the Shoulder with Loop Sling.

Grip of the Right Hand

Grasp the pistol grip with the right hand and place the forefinger on the trigger, with the thumb and remaining fingers wrapped around the pistol grip. See figure 5-14.

Figure 5-14. Grip of the Right Hand.

Firm rearward pressure should be exerted to help keep the rifle butt firmly in the shoulder, reducing the effects of recoil (this type of pressure is particularly true with the loop sling).

The trigger finger should be placed naturally on the trigger and care should be taken to ensure that the trigger finger can move independently without dragging on the side of the receiver. Proper placement of the right hand high on the pistol grip allows the trigger to be moved straight to the rear without disturbing sight alignment.

Right Elbow

The right elbow should be positioned naturally to provide balance to the position and create a pocket in the shoulder for the rifle butt. If the elbow is correctly positioned, it helps to form a pocket in the right shoulder where the rifle butt rests. The exact placement of the elbow varies with each shooting position but should remain consistent from shot to shot, ensuring the resistance to recoil remains constant. See figure 5-15.

Figure 5-15. Right Elbow.

Stock Weld

The placement of the shooter's cheek against the stock should remain firm and consistent from shot to shot. Consistency of stock weld is achieved through proper placement of the rifle butt in the pocket of the shoulder. A firm contact between the cheek and the stock enables consistent eye relief and enables the head and rifle to recoil as a single unit. Stock weld provides quick recovery between rapid fire shots, keeps the aiming eye centered in the rear sight aperture, and prevents the head from bouncing off the stock during recoil. The head should remain erect to allow the aiming eye to look straight through the rear sight aperture. See figure 5-16.

Figure 5-16. Stock Weld.

Breathing

Breathing causes chest movement and a corresponding movement in the rifle and its sights. Applying breath control will minimize this movement and the effect it has on aiming.

Muscular Tension/Relaxation

Muscular Tension–Hasty Sling

With the hasty sling donned, the shooter must apply an amount of controlled muscular tension in the left arm to keep the sling taut and stabilize the weapon sights. Resistance against the hasty sling controls the point at which the rifle sights will settle. The muscular tension is applied outward against the sling rather than in an effort to hold the rifle up. However, muscular tension should not be excessive to cause the shooter to shake, tremble, or experience fatigue.

Muscular Relaxation–Loop Sling

When using the loop sling, the muscles should be relaxed. Relaxation prevents undue muscle strain and reduces excessive movement. If proper relaxation is achieved, natural point of aim and sight alignment are more easily maintained.

5004. Elements of a Good Shooting Position

There are three elements of a good shooting position that apply when using a loop sling: bone support, muscular relaxation, and natural point of aim. The three elements of a shooting position applied with the loop sling do not apply in the same way as when firing with a hasty sling. While some degree of bone support is still achieved with the hasty sling, muscular **tension** is applied rather than muscular **relaxation**. Natural point of aim, however, applies to both the loop sling and the hasty sling.

Bone Support

The body's skeletal structure provides a stable foundation to support the rifle's weight. A weak shooting position will not withstand a rifle's repeated recoil when firing at the sustained rate or buffeting from wind. To attain a correct shooting position, the body's bones must support as much of the rifle's weight as possible. Proper use of the sling provides additional support.

The weight of the weapon should be supported by bone rather than muscle because muscles fatigue whereas bones do not.

By establishing a strong foundation for the rifle utilizing bone support, the Marine can relax as much as possible while minimizing weapon movement due to muscle tension.

Muscular Relaxation

Once bone support is achieved, muscles are relaxed. Muscular relaxation helps to hold the rifle steady and increase the accuracy of the aim. Muscular relaxation also permits the use of maximum bone support to create a minimum arc of movement and consistency in resistance to recoil. Muscular relaxation cannot be achieved without bone support. During the shooting process, the muscles of the body must be relaxed as much as possible. Muscles that are tense will cause excessive movement of the rifle, disturbing the aim. When proper bone support and muscular relaxation are achieved, the rifle will settle onto the aiming point, making it possible to apply trigger control and deliver a well-aimed shot.

Natural Point of Aim

The point at which the rifle sights settle when in a firing position is called the natural point of aim.

Since the rifle becomes an extension of the body, it may be necessary to adjust the position of the body until the rifle sights settle naturally on the desired aiming point on the target.

When in a shooting position with proper sight alignment, the position of the tip of the front sight post will indicate the natural point of aim. When completely relaxed, the tip of the front sight post should rest on the desired aiming point.

One method of checking for natural point of aim is to aim in on the target, close the eyes, take a couple of breaths, and relax as much as possible. When the eyes are opened, the tip of the front sight post should be positioned on the desired aiming point while maintaining sight alignment.

For each shooting position, specific adjustments will cause the rifle sights to settle center mass, achieving a natural point of aim.

In all positions, the natural point of aim can be adjusted by—

- Varying the placement of the left hand in relation to the handguards.
- Moving the left hand forward on the handguards to lower the muzzle of the weapon, causing the sights to settle lower on the target.
- Moving the left hand back on the handguards to raise the muzzle of the weapon, causing the sights to settle higher on the target.
- Varying the placement of the stock in the shoulder.
- Moving the stock higher in the shoulder to lower the muzzle of the weapon, causing the sights to settle lower on the target.
- Moving the stock lower in the shoulder to raise the muzzle of the weapon, causing the sights to settle higher on the target.
- Natural point of aim can be adjusted right or left by adjusting body alignment in relation to the target.

In the prone position, if the natural point of aim is above or below the desired aiming point, move the body slightly forward or back using the left elbow as a pivot and by digging the toes in.

- Pushing the body forward causes the sights to settle lower on the target.
- Pulling the body backward causes the sights to settle higher on the target.

In the kneeling and sitting positions, natural point of aim can be adjusted by varying the placement of the left elbow on the knee.

- Moving the left elbow forward on the knee lowers the muzzle of the weapon, causing the sights to settle lower on the target.
- Moving the left elbow back on the knee raises the muzzle of the weapon, causing the sights to settle higher on the target.

5005. Prone Position

Application

The prone position provides a very steady foundation for shooting and presents a low profile for maximum concealment. However, the prone position is the least mobile of the shooting positions and may restrict a Marine's field of view for observation. In this position, the Marine's weight is evenly distributed on the elbows, providing maximum support and good stability for the rifle.

Assuming the Prone Position

The prone position can be assumed by either moving forward or dropping backward into position, depending on the combat situation.

Moving Forward into Position

To move forward into the prone position, the Marine performs the following steps:

- With the hasty sling donned, stand erect and face the target, keeping the feet a comfortable distance apart (approximately shoulder width).
- Place the left hand on the handguard, the right hand on the pistol grip.
- Lower yourself into position by dropping to both knees (see fig. 5-17).

Figure 5-17. Dropping to Both Knees.

- Then shift the weight forward to lower the upper body to the ground using the right hand to break the forward motion. See figure 5-18.

Figure 5-18. Moving Forward Into Position.

Dropping Back into Position

It may be necessary to drop backward into position to avoid crowding cover, or to avoid covering uncleared terrain. To drop back into the prone position, the Marine performs the following steps:

- Face the target.
- Place the left hand on the handguard, the right hand on the pistol grip.
- Squat to the ground and break the fall with either hand (see fig. 5-19).
- Kick both legs straight to the rear (see fig. 5-20).
- If the fall was broken using the left hand, reestablish the hasty sling.

Figure 5-19. Breaking the Fall.

Figure 5-20. Kicking Back Into Position.

Straight Leg Prone Position with the Hasty Sling

Apply the seven factors to this position (para. 5003). To assume the straight leg prone position with the hasty sling, either move forward or drop back into position (see figs. 5-21, 5-22, and 5-23):

- Once on the ground, extend your left elbow in front of you. Stretch your legs out behind you. Spread the feet a comfortable distance apart with the toes pointing outboard and the inner portion of the feet in contact with the ground.
- As much of the body mass should be aligned directly behind the rifle as possible.
- If body alignment is correct, weapon recoil is absorbed by the whole body and not just the shoulder.
- Grasp the pistol grip with the right hand and place the rifle butt in the right shoulder pocket.
- Lower the head and place the cheek firmly against the stock to allow the aiming eye to look through the rear sight aperture.
- Rotate the left hand up, pinching the handguard between the thumb and forefinger.
- Slide both elbows outboard on the ground so there is outboard tension against the sling (moving the elbows out tightens the sling) and both shoulders are level. The elbows should provide a tripod of support with the body.
- Adjust the position of the left hand on the handguard to allow the sling to support the weapon and the front sight to be centered in the rear sight aperture. To adjust for elevation:
  - Move the left hand rearward or forward on the handguards (moving the hand rearward elevates the muzzle).
  - Open or close "V" of the left hand for small adjustments. (closing the "V" elevate the muzzle).
- To adjust for a minor cant in the rifle, rotate the handguard left or right in the "V" formed by the thumb and forefinger by rotating the pistol grip left or right.

Straight Leg Prone Position with the Hasty Sling

Figure 5-21. Front View.

Figure 5-22. Left View.

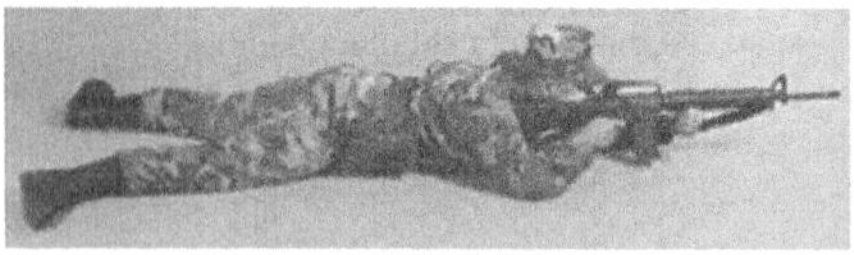

Figure 5-23. Right View.

Straight Leg Prone Position with the Loop Sling

Figure 5-24. Front View.

Figure 5-25. Left View.

Figure 5-26. Right View.

Straight Leg Prone Position with the Loop Sling

Apply the three elements and seven factors to this position (para. 5004). To assume straight leg prone position with the loop sling, either move forward or drop back into position (see figs. 5-24, 5-25, and 5-26):

- Once on the ground, roll the body to the left side as you extend and invert the left elbow on the ground. Stretch your legs out behind you. Spread the feet a comfortable distance apart with the toes pointing outboard and the inner portion of the feet in contact with the ground.
- As much of the body mass should be aligned directly behind the rifle as possible.
- If body alignment is correct, weapon recoil is absorbed by the whole body and not just the shoulder.
- Grasp the rifle butt with the right hand and place the rifle butt into the right shoulder pocket.
- Grasp the pistol grip with the right hand.
- Rotate the body to the right while the elbow is lowered to the ground so the shoulders are level. The right hand pulls and holds the rifle in the shoulder.
- Lower the head and place the cheek firmly against the stock to allow the aiming eye to look through the rear sight aperture.
- Move the left hand to a location under the handguard, which provides maximum bone support and stability for the weapon. This may require that you remove the rifle from the shoulder to reposition the left hand.

Cocked Leg Prone Position with the Hasty Sling

Figure 5-27. Front View.

Figure 5-28. Left View.

Figure 5-29. Right View.

Cocked Leg Prone Position with the Hasty Sling

Apply the seven factors to this position (para. 5003). To assume the cocked leg prone position with the hasty sling, either move forward or drop back into position (see figs. 5-27, 5-28, and 5-29):

- Once on the ground, roll the body to the left side. The left leg is stretched out behind you, almost in a straight line. This allows the mass of the body to be placed behind the rifle to aid in absorbing recoil.
- Turn the toe of the left foot inboard so the outside of the left leg and foot are in contact with the ground. Bend the right leg and draw it up toward the body to a comfortable position. Turn the right leg and foot outboard so the inside of the right boot is in contact with the ground. Cocking the leg will raise the diaphragm, making breathing easier.
- Grasp the pistol grip with the right hand and place the rifle butt in the right shoulder pocket.
- Lower the head and place the cheek firmly against the stock to allow the aiming eye to look through the rear sight aperture.
- Rotate the left hand up, pinching the handguard between the thumb and forefinger.
- Roll the body to the right while lowering the right elbow to the ground. Slide both elbows outboard on the ground so there is outboard tension against the sling (moving the elbows out tightens the sling). The right shoulder is higher than the left shoulder in the cocked leg position.
- Adjust the position of the left hand on the handguard to allow the sling to support the weapon and the front sight to be centered in the rear sight aper ture. To adjust for elevation—
  - Move the left hand rearward or forward on th handguards (moving the hand rearward elevate the muzzle).
  - Open or close "V" of the left hand for small ad justments (closing the "V" elevates the muzzle).

To adjust for a minor cant in the rifle, rotate the hand guard left or right in the "V" formed by the thumb an forefinger by rotating the pistol grip left or right.

Cocked Leg Prone Position with the Loop Sling

Apply the three elements and seven factors to this po sition (para. 5004). To assume the cocked leg pron position with the loop sling, either move forward o drop back into position (see figs 5-30, 5-31, and 5-32)

- Once on the ground, roll the body to the left sid and extend and invert the left elbow on the ground The left leg is stretched out behind you, almost in straight line. This allows the mass of the body to b placed behind the rifle to aid in absorbing recoil.
- Turn the toe of the left foot inboard so the outside o the left leg and foot are in contact with the ground Then bend the right leg and draw it up toward th body to a comfortable position. Turn the right le and foot outboard so the inside of the right boot is i contact with the ground. Cocking the leg will rais the diaphragm, making breathing easier.
- Grasp the rifle butt with the right hand and place th rifle butt into the right shoulder pocket.

Cocked Leg Prone Position with the Loop Sling

Figure 5-30. Front View.

Figure 5-31. Left View.

Figure 5-32. Right View.

- Grasp the pistol grip with the right hand.
- Roll the body to the right while lowering the right elbow to the ground. The right shoulder is higher than the left shoulder in the cocked leg position.
- Lower the head and place the cheek firmly against the stock to allow the aiming eye to look through the rear sight aperture.
- Move the left hand to a location under the handguard which provides maximum bone support and stability for the weapon.

5006. Sitting Position

There are three variations of the sitting position: crossed ankle, crossed leg, and open leg. Experiment with all the variations and select the position that provides the most stability for firing.

Although the sitting position provides an extremely stable base, it limits lateral movement and maneuver ability. It has several variations that can be adapted to the individual Marine. The sitting position provides greater elevation than the prone position while still having a fairly low profile.

Crossed Ankle Sitting Position with the Hasty Sling

The crossed ankle sitting position is an extremely stable shooting position. This position places most of the body's weight behind the weapon and aids in quick shot recovery. Apply the seven factors to this position (para. 5003). To assume the crossed ankle sitting position with the hasty sling (see figs. 5-33, 5-34, 5-35)—

- Square the body to the target.
- Grasp the handguard with the left hand.
- Bend at knees and break the fall with the right hand.
- Push backward with the feet to extend the legs and place the buttocks on the ground.
- Cross the left ankle over the right ankle.
- Grasp the pistol grip with the right hand and place the rifle butt in the right shoulder pocket.
- Lower the head and place the cheek firmly against the stock to allow the aiming eye to look through the rear sight aperture.
- Rotate the left hand up, pinching the handguard between the thumb and forefinger.
- Bend forward at the waist and place the flat portion of the back of the left arm, just above the elbow, on the left leg just below the knee.
- Place the right elbow on the inside of the right knee.
- Adjust the position to adjust sling tension. To tighten the sling: square the body more to the target, or move the left elbow out, or draw the feet up slightly toward the body.
- Adjust the position of the left hand to allow the sling to support the weapon and the front sight to be centered in the rear sight aperture.

To adjust for elevation:

- Move left hand rearward or forward on handguards (moving the hand rearward elevates the muzzle).
- Open or close the "V" of the left hand for small adjustments (closing the "V" elevates the muzzle).

To adjust for a minor cant in the rifle, rotate the handguard left or right in the "V" formed by the thumb and forefinger by rotating the pistol grip left or right.

Crossed Ankle Sitting Position with the Hasty Sling

Figure 5-33. Left View.

Figure 5-34. Front View.

Figure 5-35. Right View.

Crossed Ankle Sitting Position with the Loop Sling

Figure 5-36. Right View.

Figure 5-37. Front View.

Crossed Ankle Sitting Position with the Loop Sling

Apply the three elements and seven factors to this position (para. 5004). To assume crossed ankle sitting position with the loop sling (see figs. 5-36 and 5-37)—

- Position the body at approximately a 30-degree angle to the target.
- Place the left hand under the handguard.
- Bend at knees and break the fall with the right hand.
- Push backward with the feet to extend the legs and place the buttocks on the ground.
- Cross the left ankle over the right ankle.
- Bend forward at the waist and place the left elbow on the left leg below the knee.
- Grasp the rifle butt with the right hand and place the rifle butt into the right shoulder pocket.
- Grasp the pistol grip with the right hand.
- Lower right elbow to the inside of the right knee.
- Lower the head and place the cheek firmly against the stock to allow the aiming eye to look through the rear sight aperture.
- Move the left hand to a location under the handguard, which provides maximum bone support and stability of the weapon.

Crossed Leg Sitting Position with the Hasty Sling

The crossed leg sitting position provides a medium base of support and places some of the body's weight behind the weapon for quick recovery after each shot. Apply the seven factors to this position (para. 5003). To assume the crossed leg sitting position with the hasty sling (see figures 5-38, 5-39, and 5-40)—

- Square the body to the target.
- Grasp the handguard with the left hand.
- Bend at the knees while breaking the fall with the right hand.
- Place the buttocks on the ground as close to the feet as you comfortably can.
- Cross the left leg over the right leg.
- Grasp the pistol grip with the right hand and place the rifle butt in the right shoulder pocket.

Crossed Leg Sitting Position with the Hasty Sling

Figure 5-38. Left View.

Figure 5-39. Front

Figure 5-40. Right View.

- Lower the head and place the cheek firmly against the stock to allow the aiming eye to look through the rear sight aperture.
- Rotate the left hand up, pinching the handguard between the thumb and forefinger.
- Bend forward at the waist while placing the left elbow into the bend of the knee or placing the flat portion of the back of the left arm, just above the elbow, in front of the knee.
- Place the right elbow on the inside of the right knee.
- Adjust the position to adjust sling tension. Squaring the body more to the target or drawing the feet closer together tightens the sling by forcing the left elbow outboard.
- Adjust the position of the left hand to allow the sling to support the weapon and the front sight to be centered in the rear sight aperture.

To adjust for elevation:

- Move left hand rearward or forward on handguards (moving the hand rearward elevates the muzzle).
- Open or close the "V" of the left hand for small adjustments (closing the "V" elevates the muzzle).

To adjust for a minor cant in the rifle, rotate the handguard left or right in the "V" formed by the thumb and forefinger by rotating the pistol grip left or right.

Crossed Leg Sitting Position with the Loop Sling

Apply the three elements and seven factors to this position (para. 5004). To assume crossed leg sitting position with loop sling (see figs. 5-41, 5-42, and 5-43)—

- Position body at a 45- to 60-degree angle to target.
- Place the left hand under the handguard.
- Cross the left leg over the right leg.
- Bend at the knees while breaking the fall with the right hand.
- Place the buttocks on the ground as close to the crossed legs as you comfortably can.
- Bend forward at the waist while placing the left elbow on the left leg into the bend of the knee.
- Grasp the rifle butt with the right hand and place the rifle butt into the right shoulder pocket.
- Grasp the pistol grip with the right hand.
- Lower right elbow to the inside of the right knee.
- Lower the head and place the cheek firmly against the stock to allow the aiming eye to look through the rear sight aperture.
- Move the left hand to a location under the handguard which provides maximum bone support and stability of the weapon.

Open Leg Sitting Position with the Hasty Sling

The open leg sitting position provides a medium base of support and is most commonly used when firing from a forward slope. Apply the seven factors to this position (para. 5003). To assume the open leg sitting position with the hasty sling (see figs. 5-44, 5-45, and 5-46)—

- Square the body to the target.
- Place the feet approximately shoulder width apart.
- Grasp the handguard with the left hand.
- Bend at the knees while breaking the fall with the right hand.
- Push backward with the feet to extend the legs and place the buttocks on the ground.
- Grasp the pistol grip with the right hand and place the rifle butt in the right shoulder pocket.
- Lower the head and place the cheek firmly against the stock to allow the aiming eye to look through the rear sight aperture.

Crossed Leg Sitting Position with the Loop Sling

Figure 5-41. Left View.

Figure 5-42. Front View.

Figure 5-43. Right View.

Open Leg Sitting Position with the Hasty Sling

Figure 5-44. Left View.

Figure 5-45. Front View.

Figure 5-46. Right View.

- Rotate the left hand up, pinching the handguard between the thumb and forefinger.
- Place the flat portion of the back of the left arm, just above the elbow, in front of the left knee.
- Place the right elbow on the inside of the right knee or place the flat portion of the back of the right arm, just above the elbow, in front of the knee.
- Adjust the position to adjust sling tension. Widening the stance tightens the sling by forcing the left elbow outboard. There must be some controlled muscular tension in the legs to hold them up and offer resistance to recoil.
- The position of the left hand to allow the sling to support the weapon and the front sight to be centered in the rear sight aperture.

To adjust for elevation:

- Move left hand rearward or forward on handguards (moving the hand rearward elevates the muzzle).
- Open or close the "V" of the left hand for small adjustments (closing the "V" elevates the muzzle).

To adjust for a minor cant in the rifle, rotate the handguard left or right in the "V" formed by the thumb and forefinger by rotating the pistol grip left or right.

Open Leg Sitting Position with the Loop Sling

Apply the three elements and seven factors to this position (para. 5004). To assume the open leg sitting position with the loop sling (see figs. 5-47, 5-48, and 5-49 on page 5-16)—

- Position the body at approximately a 30-degree angle to the target.
- Place the feet approximately shoulder width apart.
- Place the left hand under the handguard.
- Bend at the knees while breaking the fall with the right hand.
- Push backward with the feet to extend the legs and place the buttocks on the ground.
- Place the left elbow on the inside of the left knee.
- Grasp the rifle butt with the right hand and place the rifle butt into the right shoulder pocket.
- Lower right elbow to the inside of the right knee.
- Lower the head and place the cheek firmly against the stock to allow the aiming eye to look through the rear sight aperture.
- Move the left hand to a location under the handguard which provides maximum bone support and stability of the weapon.

5007. Kneeling Position

Description

The kneeling position is quick to assume and easy to maneuver from. It is usually assumed after initial engagement has been made from a standing position. It can easily be adapted to available cover. A tripod is formed by the left foot, right foot, and right knee when the Marine assumes the position, providing a stable foundation for shooting. The kneeling position also presents a higher profile to facilitate a better field of view as compared to the prone and sitting positions.

Assuming the Kneeling Position

The kneeling position can be assumed by either moving forward or dropping back into position, depending on the combat situation. For example, it may be necessary to drop back into position to avoid crowding cover, or to avoid covering uncleared terrain.

Open Leg Sitting Position with the Loop Sling

Figure 5-47. Left View.

Figure 5-48. Front View.

Figure 5-49. Right View.

Moving Forward into Position

To move forward into the kneeling position, the Marine steps forward toward the target with his left foot and kneels down on his right knee.

Dropping Back into Position

To drop back into the kneeling position, the Marine leaves his left foot in place and steps backward with his right foot and kneels down on his right knee.

High Kneeling Position with the Hasty Sling

Apply the seven factors to this position (para. 5003). To assume the high kneeling position with the hasty sling, either move forward or drop back into position (see figs. 5-50, 5-51, and 5-52):

- Square the body to the target.
- Keep the right ankle straight, with the toe of the boot in contact with the ground and curled under by the weight of the body.
- Place the right portion of the buttocks on or over the right heel. Contact with the heel provides more stability to the position, however, it is not mandatory that the buttocks make contact.
- Place the left foot forward to a point that allows the shin to be vertically straight. For the shin to be vertical, the heel should be directly under the knee. The left foot must be flat on the ground. To provide a wider base of support, slide the right knee and left foot outboard to form a tripod with the right foot.
- Grasp the pistol grip with the right hand and place the rifle butt in the right shoulder pocket.
- Lower the head and place the cheek firmly against the stock to allow the aiming eye to look through the rear sight aperture.
- Rotate the left hand up, pinching the handguard between the thumb and forefinger. The left hand will not grasp the slip ring or the magazine. The magazine must be on the inside of the left arm.
- Place the flat portion of the back of the upper left arm, just above the elbow, on the left knee or against the inside of the left knee so it is in firm

High Kneeling Position with the Hasty Sling

Figure 5-50. Left View.

Figure 5-51. Front View.

Figure 5-52. Right View.

contact. The upper portion of the triceps or the armpit will not rest on the knee.

- Bend the right elbow to provide the least muscular tension possible, lowering it to a natural position.
- Adjust the position to adjust sling tension. Widening the stance by moving the left foot and knee outboard will allow the sling to be tightened.
- Adjust the position of the left hand to allow the sling to support the weapon and the front sight to be centered in the rear sight aperture.

To adjust for elevation:

- Move left hand rearward or forward on handguards (moving the hand rearward elevates the muzzle).
- Open or close the "V" of the left hand for small adjustments (closing the "V" elevates the muzzle).

To adjust for a minor cant in the rifle, rotate the handguard left or right in the "V" formed by the thumb and forefinger by rotating the pistol grip left or right.

High Kneeling Position with the Loop Sling

Apply the three elements and seven factors to this position (para. 5004). To assume the high kneeling position with the loop sling, either move forward or drop back into position (see figs. 5-53, 5-54, and 5-55):

- Position the body at a 45-degree angle to the target.
- Place the left hand under the handguard.
- Kneel down on right knee so right lower leg is approximately parallel to the target (45 to 90 degrees).
- Keep the right ankle straight, with the toe of th boot in contact with the ground and curled under b the weight of the body.
- Place the right portion of the buttocks on or over th right heel; contact with the heel provides more sta bility to the position.
- Place the left foot forward to a point that allows th shin to be vertically straight. For the shin to be ver tical, the heel should be directly under the knee The left foot must be flat on the ground since it wi be supporting the majority of the weight.
- Place the flat portion of the back of upper left arm just above the elbow, on the left knee so it is in firm contact with the flat surface formed on top of th bent knee. The point of the left elbow will exten just slightly past the left knee. The upper portion o the triceps or the armpit will not rest on the knee.
- Lean slightly forward into the sling for support.
- Grasp the rifle butt with the right hand and place th rifle butt into the right shoulder pocket.
- Grasp the pistol grip with the right hand.
- Bend the right elbow to provide the least muscula tension possible and lower it to a natural position.
- Lower the head and place the cheek firmly agains the stock to allow the aiming eye to look throug the rear sight aperture.
- Move the left hand to a location under the hand guard, which provides maximum bone support an stability for the weapon.

Medium Kneeling Position

This is also referred to as the bootless kneeling posi tion. Assume the medium kneeling position in th same way as the high kneeling position with the ex

High Kneeling Position with the Loop Sling

Figure 5-53. Left View.

Figure 5-54. Front View.

Figure 5-55. Right View.

ception of the right foot. The right ankle is straight and the foot is stretched out with the bootlaces in contact with the ground. The buttocks are in contact with the heel of the right foot. See figure 5-56.

Figure 5-56. Medium Kneeling Position.

Low Kneeling Position

The low kneeling position is most commonly used when firing from a forward slope. Assume the low kneeling position in the same way as the high kneeling position with the exception of the placement of the right foot. Turn the right ankle so the outside of the foot is in contact with the ground and the buttocks are in contact with the inside of the foot. See figure 5-57.

Figure 5-57. Low Kneeling Position.

5008. Standing Position

Description

The standing position is the quickest position to assume and the easiest to maneuver from. It allows greater mobility than other positions. The standing position is often used for immediate combat engagement. The standing position is supported by the shooter's legs and feet and provides a small area of contact with the ground. In addition, the body's center of gravity is high above the ground. Therefore, maintaining balance is critical in this position.

Standing Position with the Hasty Sling

Apply the seven factors to this position (para. 5003). To assume the standing position with the hasty sling (see figs. 5-58, 5-59, and 5-60)—

- Square the body to the target.
- Spread the feet apart to a comfortable distance with the left foot slightly in front of the right foot. This distance may be wider than shoulder width.
- Distribute the weight evenly over both feet and hips. Balance will shift forward slightly to reduce recovery time and increase the stability of the hold. The legs should be slightly bent for balance.
- Grasp the pistol grip with the right hand and place the rifle butt in the right shoulder pocket.
- Bring the rifle sights up to eye level instead of lowering the head to the sights and place the cheek firmly against the stock. Ensure the head is erect so the aiming eye can look through rear sight aperture.
- Rotate the left hand up, pinching the handguard between the thumb and forefinger.
- The left hand will be under the handguard with the thumb on the outboard side of the handguard.The left hand will not grasp the slip ring or magazine. The magazine must be on the inside of the left arm.
- Hold the right elbow in a natural position.
- Adjust the position of the left hand on the handguard to allow the sling to support the weapon and front sight to be centered in the rear sight aperture.

To adjust for elevation:

- Move left hand rearward or forward on handguards (moving the hand rearward elevates the muzzle).
- Open or close the "V" of the left hand for small adjustments (closing the "V" elevates the muzzle).
- Adjust the position to adjust sling tension. Moving the left elbow out or squaring the body more to the target tightens the sling.

To adjust for a minor cant in the rifle, rotate the handguard left or right in the "V" formed by the thumb and forefinger by rotating the pistol grip left or right.

Standing Position with the Hasty Sling

Figure 5-58. Left View.

Figure 5-59. Front View.

Figure 5-60. Right View.

Standing Position with the Parade Sling

The parade sling is used to emphasize marksmanship fundamentals while firing from the standing position on KD courses during entry-level training. To achieve proper positioning of the parade sling, perform the following steps:

- Attach sling to rifle by placing feed end of sling down through the upper sling swivel.
- Place feed end of sling through sling keeper and lock into place.
- Attach J-hook to lower sling swivel.
- Pull feed end of sling through sling keeper until sling is taut.
- Move sling keeper down near feed end of sling.
- Lock sling keeper into place.

Perform the following steps to assume the standing position with a parade sling (figure 5-61):

- Stand erect.
- Face approximately 90 degrees to the right of the line of fire.
- Place feet approximately shoulder width apart.
- Place left hand under handguard.
- Grasp pistol grip with right hand.
- Place rifle butt into right shoulder pocket.
- Invert left elbow across rib cage.
- Rest left arm naturally against rib cage.
- Lower right elbow to a natural position.
- Place cheek firmly against stock to obtain a firm stock weld.

Figure 5-61. Standing Position with the Parade Sling.

CHAPTER 6. USE OF COVER AND CONCEALMENT

Note

+The procedures in this manual are written for right-handed Marines; left-handed Marines should reverse instructions as necessary.

6001. Cover and Concealment

In a combat environment, a Marine must be prepared to fire from any type of cover or concealment. Cover is anything that protects a Marine from enemy fire. Cover may be an existing hole, a hastily dug shelter, or a well-prepared fighting position with overhead protection. Concealment is anything that hides a Marine from enemy view, but it may not afford protection. Concealment can be obtained from buildings, trees, crops, and skillful use of ground contours. A Marine can use any object or terrain feature that protects him from enemy fire, hides him from enemy view, allows him to observe the enemy, and provides support for a firing position.

Types of Cover

Frontal Cover

A firing position should have frontal cover that provides protection from small arms fire and indirect fire fragments. Ideally, frontal cover should be thick enough to stop small arms fire, high enough to provide protection from enemy fire, and wide enough to provide cover when firing to the left or right edge of a sector of fire.

Ideal Cover

The ideal cover provides:

- Overhead, flank, and rear protection from direct and indirect fire.
- Free use of personal weapons.
- Concealment from enemy observation.
- A concealed route in and out.
- Unobstructed view of the sector of fire.

Common Cover Materials

Any material that protects a Marine from small arms fire can be used for cover. Some common materials include sandbags, trees, logs, and building debris. Table 6-1 presents some common materials and their minimum thickness required for protection from small arms fire.

Table 6-1. Minimum Thickness for Protection Against Small Arms.

Material	Minimum Thickness (in inches)
Concrete	7
Broken stone (rubble)	20
Dry sand	24
Wet sand	35
Logs (oak)	40
Earth (packed)	48

Sandbags

Cover can be improved and positions can be fortified by filling sandbags with dirt/sand and placing them around the position. Sandbags should be tightly packed because bullets can easily penetrate moist or loosely packed sandbags. Overlapping sandbags increase protection and decrease the bullet's ability to penetrate the sandbag. A minimum thickness of three sandbags is required to stop small arms fire.

Trees/Logs

Wood is a relatively dense material and offers good cover and protection. Bullets have a tendency to fragment when they penetrate wood. Live trees have a greater resistance to bullet penetration than dead trees. Wood that has been treated with creosote, such as telephone poles and railroad ties, offers better protection from projectiles than untreated wood, but it still does not ensure protection from small arms fire.

Cinder Bocks

Cinderblocks are not impenetrable cover. Although they are made of a dense material, the composition of a cinderblock is so brittle that a bullet can shatter the block upon impact. This can cause injury to a Marine by secondary fragmentation.

Firing From Specific Types of Cover

Effective cover allows a Marine to engage enemy targets while protecting himself from enemy fire. Several types of cover provide support, protection, and concealment and do not interfere with target engagement. A Marine must adapt firing positions to the type of cover available.

Fighting Hole

A Marine should use fighting holes if available. See figure 6-1. After a Marine enters the fighting hole, he adds or removes dirt, sandbags or other supports to fit his height. To assume a firing position, a Marine performs the following steps:

- Place the right foot to the rear as a brace.
- Lean forward until the chest is against the forward wall of the fighting hole.
- Extend the left arm and elbow over the forward side of the fighting hole so the left forearm rests against the back of the parapet.
- Place the rifle butt into the pocket of the right shoulder and grasp the pistol grip with the right hand.
- Place the right elbow on solid support using the elbow rest of the fighting hole or sandbags placed around the fighting hole.

Figure 6-1. Fighting Hole Position.

Rooftop

If possible, a Marine's entire body should be positioned behind the apex of the rooftop, using the apex to support the rifle. See figure 6-2. If the body cannot be positioned behind the apex, place the left arm over the apex of the roof to hold the weight of the body. Expose as little of the head and shoulders as possible. See figure 6-3.

Figure 6-2. Rooftop Position Behind the Apex.

Figure 6-3. Rooftop Position Supported by the Apex.

Window

The Marine can establish a supported or unsupported position from a window.

Unsupported. A Marine can establish an unsupported position back from the opening of the window so that the muzzle does not protrude and interior shadows provide concealment so as not to provide a silhouette to the enemy. See figure 6-4.

Figure 6-4. Unsupported Window Position.

Supported. When additional stability is needed, a Marine can establish a supported position by placing the rifle handguards or his forearm in the "V" formed by the side and bottom of the windowsill. A drawback to this technique is that the muzzle of the weapon and the Marine may be exposed to view. See figures 6-5 and 6-6.

Vehicle

In many combat situations, particularly in urban environments, a vehicle may be the best form of cover. When using a vehicle for cover, the engine block provides the most protection from small arms fire. The Marine should establish a position behind the front wheel so the engine block is between him and the target (see figs. 6-7 and 6-8). From this position, the Marine may fire over, under or around the vehicle. This is a particularly effective position for larger vehicles that are high off the ground.

The Marine can establish additional support for the rifle by positioning himself behind the doorjamb (frame of door) and placing the rifle against the "V" formed by the open door and doorframe (see figure 6-9). From this position, the Marine may fire over the hood of the vehicle while using the engine block for protection.

Figure 6-5. Supported Window Position (Handguards on Support).

Figure 6-6. Supported Window Position (Forearm on Support).

Figure 6-7. Firing Around Front of Vehicle.

Figure 6-8. Firing Over Front of Vehicle.

Figure 6-9. Establishing a Supported Position in a Vehicle.

However, this position limits lateral mobility and it is more difficult to maneuver from.

At the back of the vehicle, only the axle and the wheel provide cover. If the Marine must shoot from the back of the vehicle, he must position himself directly behind the wheel as much as possible (see fig. 6-10).

Figure 6-10. Using the Back of a Vehicle for Cover.

6002. Supported Firing Positions

Supports are foundations for positions; positions are foundations for the rifle. To maximize the support the position provides, the firing position should be adjusted to fit or conform to the shape of the cover. Elements of a sound firing position, such as balance and stability, must be incorporated and adjusted to fit the situation and type of cover. A supported firing position should minimize exposure to the enemy, maximize the stability of the rifle, and provide protection from enemy observation and fires. A Marine can use any available support (e.g., logs, rocks, sandbags or walls) to stabilize his firing position. The surrounding combat environment dictates the type of support and position used.

Considerations Using Cover and Concealment

Cover and concealment considerations are similar, regardless of the combat environment (e.g., urban, desert, jungle).

Adjusting the Shooting Position

The type of cover can dictate which shooting position (e.g., standing, kneeling, sitting, and prone) will be the most effective. For example, a Marine's height in relation to the height of the cover aids in the selection of a firing position.

The firing position selected should be adjusted to fit the type of cover to:

- Provide stability. The position should be adjusted to stabilize the rifle sights and allow the management of recoil to recover on target.
- Permit mobility. The position should be adjusted to permit lateral engagement of dispersed targets and movement to other cover.
- Allow observation of the area/enemy while minimizing exposure to the enemy.

The firing position is adjusted to fit the type of cover by adjusting the seven factors (i.e., left hand, pocket of shoulder, right elbow, stock weld, grip of right hand) to support the rifle or the position.

Keeping the Entire Body Behind Cover

A Marine should minimize exposure of any part of his body to fire. Be especially aware of the head, right elbow, knees or any other body part that may extend beyond the cover.

Firing From the Right or Left Side of Cover

To minimize exposure and maximize the cover's protection, a right-handed Marine should fire from the right side of cover and a left-handed Marine should fire from the left side, if possible (see fig. 6-11).

Figure 6-11. Firing from the Right Side of Cover.

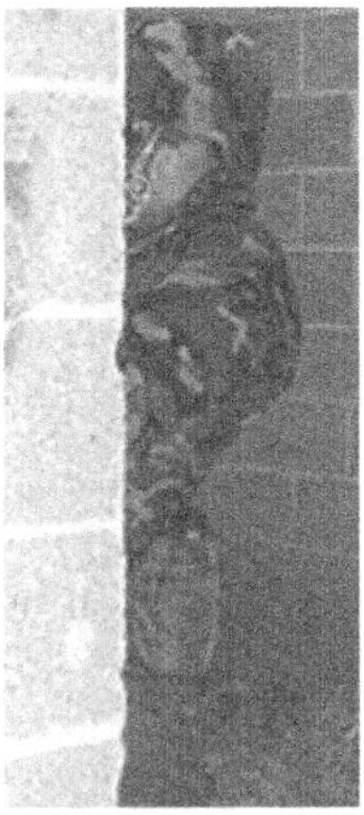

Figure 6-12. Firing from the Left Side of Cover.

If, however, a right-handed Marine must fire from the left side of cover, he fires right-handed but adjusts his position behind cover and uses the rollout technique (see para. 6003) to engage the target. See figure 6-12.

Firing Over the Top of Cover

Firing over the top of cover provides a wider field of view and lateral movement. When firing over the top of cover, the position may be supported and stabilized by resting the handguard or the left forearm on the cover (see fig. 6-13). The Marine should keep as low a profile as possible; the rifle should be as close to the top of cover as possible.

Figure 6-13. Firing Over the Top of Cover.

Maintaining Muzzle Awareness

When firing over the top of cover, a Marine must remember that the sights are higher than the barrel and remain aware of the location of his muzzle. Therefore, a Marine must maintain a position that ensures the muzzle is high enough to clear the cover (e.g., window sill, top of wall) as he obtains sight alignment/sight picture on the target (see figs. 6-14a and 6-14b).

Clearing the Ejection Port

Ensure the cover does not obstruct the ejection port. If the ejection port is blocked, the obstruction can interfere with the ejection of the spent cartridge case and cause a stoppage.

Figures 6-14a and 6-14b. Clearing Cover with the Muzzle.

Resting the Magazine

The bottom, front or side of the rifle magazine can rest on or against support to provide additional stability (see figs. 6-15, 6-16, and 6-17).

CAUTION

The back of the magazine should not be pulled back against support because it can cause a stoppage by not allowing a round to feed from the magazine.

Seven Factors

The seven factors (left hand, rifle in shoulder pocket, stock weld, right elbow, grip of the right hand, breathing, muscular tension) are applied when firing from cover, however, some may have to be modified slightly to accommodate the artificial support provided the rifle and position.

Left Hand

The support should be used to help stabilize both the firing position and the rifle to enable the Marine to maintain sight alignment and sight picture.

The forearm or left hand can contact the support to stabilize the weapon. Rest the forearm or the meaty portion of the bottom of the left hand on the support and rest the rifle in the "V" formed by the thumb and forefinger of the left hand (see fig. 6-18).

The rifle's handguards may rest on the support, but the barrel may not (see fig. 6-19). Placement of left hand on the handguard may have to be adjusted forward or back to accommodate the cover and the additional support provided by the rifle resting on the cover. If the handguards are resting on the cover, the left hand

Figure 6-15. Bottom of Magazine on Support.

Figure 6-16. Front of Magazine on Support.

Figure 6-17. Side of Magazine on Support.

can pull down on the handguards to further stabilize the weapon.

Rifle in Shoulder Pocket

Regardless of the combat situation or the height of the cover, the rifle butt must remain in the shoulder pocket to manage the effects of recoil, stabilize the rifle, and maintain the rifle's battlesight zero. The firing position must be adjusted behind cover to enable the rifle to be placed in the shoulder.

Stock Weld

Regardless of the cover, the firing position must be adjusted to allow stock weld to be achieved. Proper stock weld provides quick recovery between shots and keeps the aiming eye centered in the rear sight aperture.

Right Elbow

The right elbow can be placed on or against the support to stabilize the weapon and the position. Ensure the elbow is not extended beyond cover revealing the position to the enemy.

Grip of the Right Hand

If rifle handguards, forward hand, or forearm rest on cover for support, the grip of the right hand should pull back and down on the pistol grip to further stabilize the weapon in the shoulder and on the support.

Breathing

Breathing does not change when firing from a supported position.

Muscular Tension

To create balance and support for the position, the Marine may shift his body weight into or against the support (see fig. 6-20). This enables the Marine to use cover to support his body weight, reducing the need for muscular tension.

Figure 6-18. Forearm Resting on Cover.

Figure 6-19. Handguards Resting on Cover.

Figure 6-20. Shifting Body Weight into Cover.

Types of Supported Positions

Supported Prone

If possible, a Marine should use the supported prone position when firing from behind cover. This position is the steadiest, provides the lowest silhouette, and provides maximum protection from enemy fire.

- Support the position by placing the handguards, the forearm, or the magazine on or against support (see fig. 6-21).

Figure 6-21. Supported Prone.

- The prone position can be assumed behind a tree, a wall, a log, or almost any type of cover. It is flexible and allows shooting from all sides of cover and from cover of various sizes.
- The body must be adjusted to conform to the cover. For example, if the cover is narrow, keep the legs together. The body should be in line with the rifle and directly behind the rifle (see figs. 6-22 and 6-23). This presents a smaller target to the enemy and provides more body mass to absorb recoil.

Figure 6-22. Supported Prone Behind Narrow Cover—Rear View.

Supported Kneeling

When the prone position cannot be used because of the height of the support, the supported kneeling position may be appropriate. The kneeling position provides additional mobility over the prone position.

- The kneeling position allows shooting from all sides and from cover of varying sizes. This position may be altered to maximize the use of cover or support by assuming a variation of the kneeling position (high, medium or low).
- In the kneeling position, the Marine must not telegraph his position behind the cover with his knee. When shooting around the sides of cover, the Marine should strive to keep his right knee in line with his left foot so as not to reveal the position to the enemy. See figure 6-24.
- Support the position by placing the handguards, the forearm, or the magazine on or against support. In addition, the position (e.g., a knee, the side of the body) may rest against support (see fig. 6-25).
- If the rifle is resting on support, the Marine may not need to stabilize the weapon by placing his left elbow on his knee (see fig. 6-26).

Supported Sitting

A supported sitting position may be used to fire over the top of cover when mobility is not as critical. A sitting position can be comfortably assumed for a longer period of time than a kneeling position and it can conform to higher cover when a prone position cannot be used.

Figure 6-23. Supported Prone Behind Narrow Cover—Front View.

Figure 6-24. Supported Kneeling.

Figure 6-25. Supported Kneeling with Body Against Support.

- Support the position by placing the handguards, the forearm, or the magazine on or against support (see fig. 6-27).

- If the rifle is resting on support, the Marine may not need to stabilize the weapon by placing his left or right elbows on his legs (see fig. 6-28).

Supported Standing

The supported standing position provides greater mobility than the other positions and usually provides greater observation of the enemy. The supported standing position is effectively used behind high cover (e.g., window, over a wall) or narrow cover (e.g., tree, telephone pole).

- To assume the supported standing position, the Marine leans his body forward or against support to stabilize the weapon and the position.

Figure 6-26. Supported Kneeling with Rifle Resting on Support.

Figure 6-27. Supported Sitting.

Figure 6-28. Supported Sitting with Rifle on Support.

- Support the position by placing the handguards, the forearm, or the magazine on or against support. In addition, the position (e.g., the side of the body) may rest against support (see fig. 6-29).

Figure 6-29. Supported Standing.

6003. Searching for and Engaging Targets From Behind Cover

To locate targets when behind cover or to ensure the area is clear before moving, the Marine must expose as little of himself as possible to the enemy. Additionally, the Marine must be ready to fire if a target is located. Two techniques that can be used to locate and engage targets from behind cover are: the pie and rollout. These techniques minimize the Marine's exposure to enemy fire while placing the Marine in a position to engage targets or to move to another location if necessary. These techniques are also used to enter a building or structure. Both techniques are used in the kneeling and standing positions. To be accurate in engaging targets using either technique, the seven factors must be applied. See page 6-6 for a discussion of seven factors.

Pie Technique

To perform the pie technique:

- Staying behind cover, move back and away from the leading edge of the cover. The surroundings and situation will dictate the distance you should move back and away from the cover. Generally, the further back the Marine is from cover, the greater his area of observation; staying too close to cover decreases the area of observation. However, if the Marine is too far back from the leading edge of cover, he may unknowingly expose himself to the enemy.
- Assume a firing position and lower the rifle sights enough to have a clear field of view, orienting the muzzle on the leading edge of the cover. (In a building, the baseboards serve as a reference point for the muzzle of the rifle when searching for targets.)
- Taking small side steps, slowly move out from behind the cover, covering the field of view with the aiming eye and muzzle of the weapon. Wherever the eyes move, the muzzle should move (eyes, muzzle, target). The muzzle should remain on the leading edge of cover, serving as a pivot point when moving out.
- Continue taking small side steps and moving out from cover until a target is identified or the area is found to be clear.
- When a target is identified, sweep the safety, place the finger on the trigger, and engage the target.

Rollout Technique

To perform the rollout technique:

- Staying behind cover, move back and position the body so it is in line with the leading edge of the cover, ensuring that no part of the body extends beyond the cover.
- Assume a firing position and come to the Ready, ensuring the muzzle is just behind the cover.
- Canting the head and weapon slightly, roll the upper body out to the side just enough to have a clear field of view and allow the muzzle to clear the cover. Keeping the feet in place, push up on the ball of one foot to facilitate rolling out.
- Continue rolling out from cover until a target is identified or the area is found to be clear.
- When a target is identified, sweep the safety, place the finger on the trigger, and engage the target.
- If a target is identified before moving out from cover, the rifle should be taken off safe before moving out using the rollout technique.

Combining Techniques

In some situations, it may be necessary to combine the pie and rollout techniques to search an entire area for targets (e.g., corner of a building, a doorway). Changing from one technique to another may permit the Marine to minimize his exposure to enemy observation and fires.

6004. Moving Out From Behind Cover

A Marine must be constantly aware of his surroundings and available cover, should a threat appear. He should avoid obvious danger areas and move quickly through danger areas that cannot be avoided.

When moving from cover to cover, the Marine should select the next covered location and plan his route before moving from his present position. This is done by quickly looking from behind cover to ensure the area is clear, ensuring the head and eyes are exposed for as short a time as possible.

If necessary, the Marine should conduct a Condition 1 reload before moving from cover.

Once the Marine is committed to moving, all focus should be on moving until cover is reassumed.

Chapter 7. Rifle Presentation

Note

☞The procedures in this manual are written for right-handed Marines; left-handed Marines should reverse instructions as necessary.

7001. Presentation of the Rifle

In a combat environment, targets may present themselves with little or no warning. To maintain an advantage, the Marine carries his weapon in a position appropriate to the threat level that permits the rifle to be both easily carried and presented as quickly as possible. A carry is also established based on the situation such as moving in a close quarter environment, moving over or under objects, etc.

Presenting the Rifle From the Tactical Carry

The Marine uses the Tactical Carry when no immediate threat is present. This carry permits the rifle to be easily carried for long periods of time, but it does not permit the quickest presentation to a target. If the situation changes and a target presents itself, a Marine performs the following steps to present the rifle from the Tactical Carry once a target appears:

- Extend the rifle toward the target keeping the muzzle slightly up so the buttstock clears all personal equipment. Continue to look at the target.
- At the same time, place the rifle in Condition 1. Two methods can be used to place rifle in Condition 1 if it is in Condition 3:
  - Grip the pistol grip firmly with the right hand. Pull the charging handle with the left hand to its rearmost position and release (see fig. 7-1).

Figure 7-1. Pulling Charging Handle with Left Hand.

  - Grip the handguards firmly with the left hand. Pull the charging handle with the right hand to its rearmost position and release (see fig. 7-2).

Figure 7-2. Pulling Charging Handle with Right Hand.

- As the rifle is being presented, take the rifle off safe and place the trigger finger on the trigger (see figs. 7-3a, 7-3b, and 7-3c on page 7-2).
- Level the rifle while pulling it firmly into the pocket of the shoulder to obtain proper stock weld. Do not move the head down to meet the stock of the rifle.

Note

If the rifle is in the shoulder properly, the aiming eye will be able to look through the rear sight as soon as the stock makes contact with the cheek.

- As the sights become level with the aiming eye, visually locate the target through the rear sight aperture. As the rifle sights settle, shift the focus back to the front sight post to obtain sight alignment, and place the tip of the post center mass on the target to obtain sight picture.

Presenting the Rifle From the Alert Carry and From the Ready Carry

The Marine uses the Alert Carry when enemy contact is likely. The Alert is also used for moving in close terrain (e.g., urban, jungle).

The Marine uses the Ready Carry when enemy contact is imminent.

Figures 7-3a, 7-3b, and 7-3c. Sweeping the Safety.

To present the rifle from the Alert and from the Ready, a Marine performs the following steps once a target appears:

- While looking at the target, bring the muzzle up by raising the left hand, allowing the rifle butt to pivot in the shoulder. At the same time, pull the rifle firmly into the pocket of the shoulder.
- As the rifle is being presented, take the rifle off safe and place the trigger finger on the trigger (see figs. 7-3a, 7-3b, and 7-3c).
- As the stock makes contact with the cheek, level the rifle to obtain proper stock weld.
- Do not move the head down to meet rifle stock.

Note

If the rifle is in the shoulder properly, the aiming eye will be able to look through the rear sight as soon as the stock makes contact with the cheek.

- As the sights become level with the aiming eye, visually locate the target through the rear sight aperture. As the rifle sights settle, shift the focus back to the front sight post to obtain sight alignment, and place the tip of the post center mass on the target to obtain sight picture.

Presenting the Rifle From the Strong Side Sling Arms Transport

Once a target appears, a Marine performs the following to present the rifle from Strong Side Sling Arms:

- While looking at the target, lean forward slightly to facilitate removal of the rifle from the shoulder.
- Reach under the right arm with the left hand between the sling and the body and grasp the handguards. See figure 7-4). At the same time, pull down on the sling and raise right elbow out and parallel to the deck.

Figure 7-4. Grasping the Handguards.

- Roll the right shoulder forward and release the sling from the right hand once the handguards have cleared the elbow. At the same time, pull rifle forward off the shoulder with left hand.
- Continue pulling the rifle forward with the left hand while rotating the rifle parallel to the deck; when the right arm is free of the sling and the rifle clears all personal gear, grasp the charging handle with the right hand (see fig. 7-5) and pull it to its rearmost position and release.
- Establish a firing grip with right hand while keeping the trigger finger straight along the receiver.
- Take rifle off safe and place the trigger finger on the trigger (see figs. 7-3a, 7-3b, and 7-3c).
- Level the rifle while pulling it firmly into the pocket of the shoulder to obtain proper stock weld. Do not move the head down to meet the stock of the rifle.

Figure 7-5. Clearing Gear and Grasping the Charging Handle.

Note

If the rifle is in the shoulder properly, the aiming eye will be able to look through the rear sight as soon as the stock makes contact with the cheek.

- As the sights become level with the aiming eye, visually locate the target through the rear sight aperture. As the rifle sights settle, shift the focus back to the front sight post to obtain sight alignment, and place the tip of the post center mass on the target to obtain sight picture.

Presenting the Rifle From the Weak Side Sling Arms Transport

The hasty sling should be maintained while presenting the rifle from this transport. To present the rifle from Weak Side Sling Arms, a Marine performs the following steps once a target appears:

- While looking at the target, lean forward slightly to facilitate removal of rifle from the shoulder.
- Grasp the sling with the right hand to prevent the rifle from falling off the shoulder.
- Grasp handguards with left hand (the index finger points toward the muzzle). See figure 7-6.
- Rotate the rifle counterclockwise while extending the muzzle toward the target.
- Continue extending the rifle toward the target to ensure the rifle clears all personal gear.
- Grasp the charging handle with the right hand and pull it to its rearmost position and release (see fig. 7-7).
- Establish a firing grip with right hand while keeping the trigger finger straight along the receiver.
- Take the rifle off safe and place the trigger finger on the trigger (see figures 7-3a, 7-3b, and 7-3c).
- Level the rifle while pulling it firmly into the pocket of the shoulder to obtain proper stock weld. Do not move the head down to meet the stock of the rifle.

Figure 7-6. Grasping the Handguards.

Note

If the rifle is in the shoulder properly, the aiming eye will be able to look through the rear sight as soon as the stock makes contact with the cheek.

Figure 7-7. Clearing Gear and Grasping the Charging Handle.

- As the sights become level with the aiming eye, visually locate the target through the rear sight aperture. As the rifle sights settle, shift the focus back to the front sight post to obtain sight alignment, and place the tip of the post center mass on the target to obtain sight picture.

7002. Search and Assess

After a Marine engages a target, he must immediately search the area and assess the results of his engagement. Searching and assessing enables the Marine to avoid tunnel vision that can restrict the focus so that an indication of other targets is overlooked.

Purpose

The Marine searches the area for additional targets or for cover. The Marine assesses the situation to determine if he needs to re-engage a target, engage a new target, take cover, assume a more stable position, cease engagement, etc.

Technique

To search and assess, a Marine performs the following steps:

- Keeping the buttstock in the shoulder, lower the muzzle of the rifle slightly to look over the sights.
- Place the trigger finger straight along the receiver (see fig. 7-8).
- Search the area and assess the situation/threat by moving the head, eyes, and rifle left and right (approximately 45 degrees from center) to cover the immediate area. The muzzle moves with the head and eyes in one fluid motion while searching. Keep both eyes open to increase the field of view.

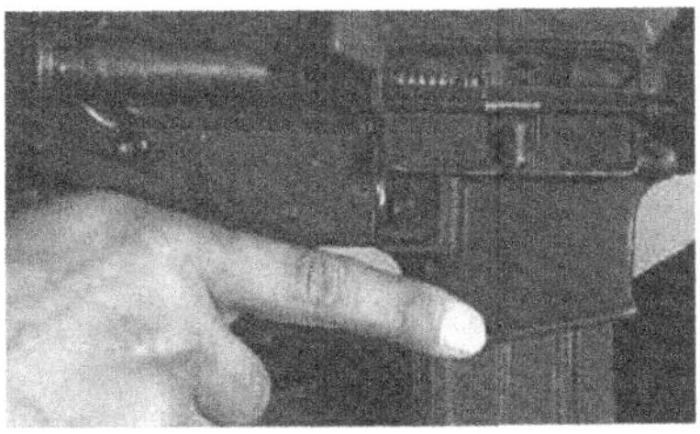

Figure 7-8. Straight Trigger Finger.

- Once a Marine determines the area is clear of enemy threat, he places the rifle on safe.

Searching and Assessing to a Higher Profile

Depending on the tactical situation, the Marine may choose to increase his area of observation by searching and assessing to a higher profile position.

Prone to Kneeling

After searching and assessing at the prone position, move to a kneeling position by performing the following steps:

- While maintaining control of the pistol grip, lower the rifle butt out of the shoulder.
- Drop the left hand to the deck and, bringing it back, push up off the deck to both knees (see figs. 7-9a and 7-9b).
- Grasp the handguard with the left hand and place the rifle butt in the pocket of the shoulder.
- Assume a kneeling position and search and assess (see fig. 7-10).

Sitting to Kneeling

After searching and assessing at the sitting position, move to a kneeling position by performing the following steps:

- Maintain control of the rifle with the rifle butt in the pocket of the shoulder.

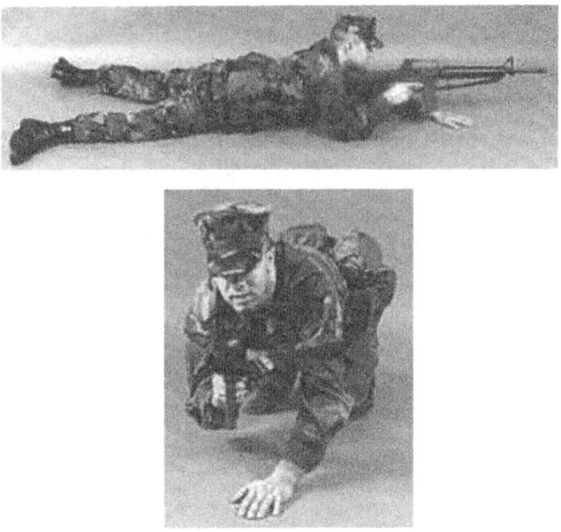

Figures 7-9a and 7-9b. Pushing Up Off the Deck to Both Knees.

- Uncross the legs to an open leg position.
- Tuck the right foot underneath the left thigh, as close to the buttocks as possible (see fig. 7-11).
- Lean forward and to the right and roll on to the right knee to a kneeling position and search and assess (see fig. 7-12).

Note

It may be necessary to release the rifle with the right hand and push off the deck with the right hand to assist in rolling up to a kneeling position.

Kneeling to Standing

After searching and assessing at the kneeling position maintain control of the rifle with the rifle butt in the pocket of the shoulder, and stand while continuing to search and assess.

Figure 7-10. Kneeling Search and Assess.

Figure 7-11. Tucking the Right Foot.

Figure 7-12. Rolling Up to a Kneeling Position.

Chapter 8. Effects of Weather

Wind, temperature, and precipitation can affect the trajectory of the bullet. In addition, all weather conditions have a physical and psychological effect on Marines. Marines must use techniques to offset the effects of wind, light, temperature, and precipitation (snow, sleet, and rain). Through proper training, Marines can develop the confidence required to reduce the physical and psychological effects of weather during combat situations.

8001. Physical Effects of Wind on the Bullet

Physical Effects

The weather condition that presents the greatest problem to shooting is the wind. Wind affects a bullet's trajectory. The effect of wind on the bullet as it travels down range is referred to as deflection. The wind deflects the bullet laterally in its flight to the target (see fig. 8-1).

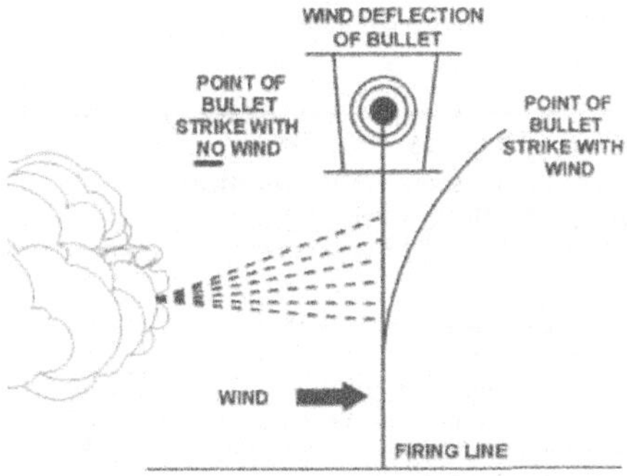

Figure 8-1. Deflection of a Bullet.

The bullet's exposure time to the wind determines the amount the bullet is deflected from its original trajectory. Deflection increases as the distance to the target increases. There are three factors that affect the amount of deflection of the bullet:

- Velocity of the wind—The greater the velocity of the wind, the more the bullet will be deflected.
- Range to the target—As the distance to the target increases, the speed of the bullet slows allowing the wind to have a greater effect on shot placement.
- Velocity of the bullet—A bullet with a high muzzle velocity will not be affected by the wind as much as a bullet with a low muzzle velocity.

Determining Windage Adjustments to Offset Wind Effects

The velocity and direction of the wind in relationship to the bullet must be determined to offset the wind's effects. If Marines can classify wind values and determine velocity within 5 mph, they can effectively engage targets in windy conditions.

Wind Direction

Determine wind direction by observing direction vegetation is moving, by feeling the wind blow against the body, or by observing direction of a flag (in training).

Wind Value Classifications

Winds are classified according to the direction from which they are blowing in relation to the direction of fire. The clock system indicates wind direction and value (see fig. 8-2). Winds can be classified as half

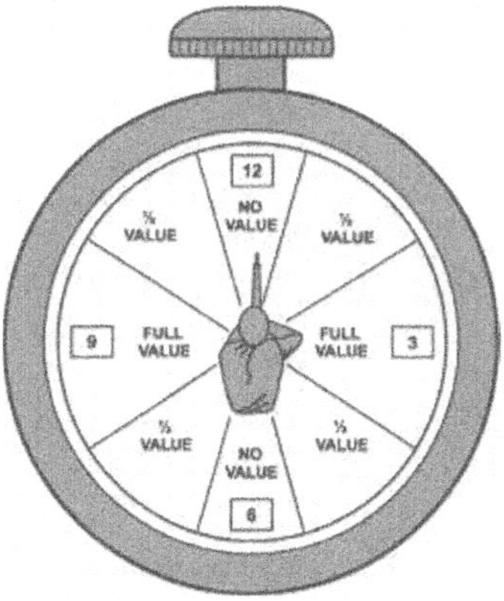

Figure 8-2. Clock System.

value, full value, or no value. The target is always located at 12 o'clock.

Wind Velocity

There are two methods used to determine wind velocity: observation and flag. The flag method is used as a training tool on the known distance (KD) range to learn the observation method. This method teaches Marines to relate the effect a given wind condition has on the natural surroundings in order to develop the base of knowledge used during the observation method. The observation method is the primary method used to estimate wind velocity and direction in a tactical situation. The following are guidelines used during the observation method:

- Under 3 miles per hour (mph) the wind can hardly be felt on the face. The presence of a slight wind can be detected by drifting smoke.
- 3 to 5 mph winds can be felt lightly on the face.
- 5 to 8 mph winds keep tree leaves in a constant motion.
- 8 to 12 mph winds raise dust and loose paper.
- 12 to 15 mph winds cause small trees to sway.
- 16 to 25 mph winds cause large trees to sway.

Flag Method

The flag method is primary method used on the KD range. To estimate wind velocity in miles per hour:

Estimate the angle created between the flagpole and the flag in degrees.

- Divide the angle by four to estimate wind velocity in miles per hour. See figure 8-3.

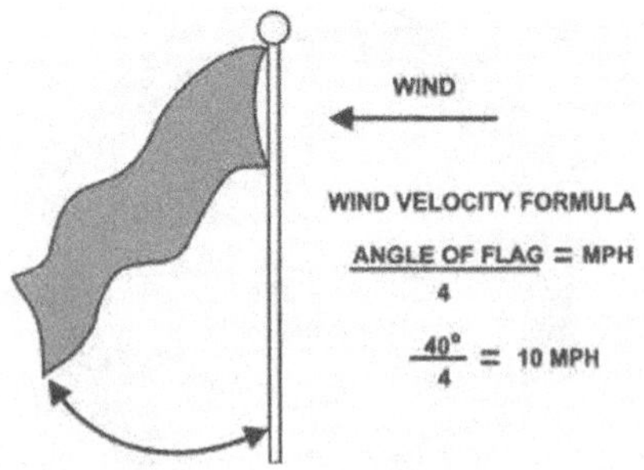

Figure 8-3. Flag Method.

Note

Information given is based on a dry flag. A wet flag is heavy and gives a false reading.

Windage Adjustments

After identifying wind direction, wind classification, and wind velocity, windage adjustments needed to enable the bullet to strike the target are estimated in the following ways:

Observation Method. Using the windage chart provided in figure 8-4, match the wind velocity, wind direction, and range to the target to the information in the chart to estimate the correct number of clicks to apply to the windage knob.

Flag Method. Using the windage chart provided in figure 8-5, match the wind velocity, wind direction, and range to the target to the information in the chart to determine the correct number of clicks to apply to the windage knob.

Once the number of windage clicks is determined, turn the windage knob causing the rear sight aperture to move into the direction of the wind. (See chapter 9.)

8002. Physical Effects of Temperature and Precipitation on the Bullet and the Rifle

Temperature

Extreme changes in temperature cause fluctuation in the rifle's chamber pressure. This fluctuation is caused by changes in the propellant's temperature. In cold weather, as rifle chamber pressure decreases, the bullet exits the muzzle at a lower velocity, and the bullet impacts the target below the point of aim. In extreme heat, the rifle's chamber pressure increases causing the bullet to exit the muzzle at a higher velocity and impact the target above the point of aim. Hot air is less dense than cool air and provides less resistance to the bullet; this allows the bullet to travel faster and experience less deflection from the wind. Cold air is dense and provides the bullet with more resistance; this causes the bullet to travel slower and experience greater deflection from the wind.

RANGE YARDS	WINDS CAN BE FELT LIGHTLY ON FACE AND LEAVES ARE IN CONSTANT MOTION		WINDS RAISE DUST AND LOOSE PAPER 8-12 MPH		WINDS CAUSE SMALL TREE TO SWAY 12-15 MPH		WINDS CAUSE LARGE TREES TO SWAY 16-20 MPH		21-25 MPH	
	WIND VALUE		WIND VALUE		WIND VALUE		WIND VALUE		WIND VALUE	
	FULL	HALF	FULL	HALF	FULL	HALF	FULL	HALF	FULL	HALF
200	2	1	3	1	5	2	6	3	8	4
300	3	1	6	3	10	5	13	6	16	8
400	4	2	9	4	13	6	18	9	22	11
500	6	3	12	6	18	9	24	12	30	15

Figure 8-4. Windage Click Chart for the Observation Method.

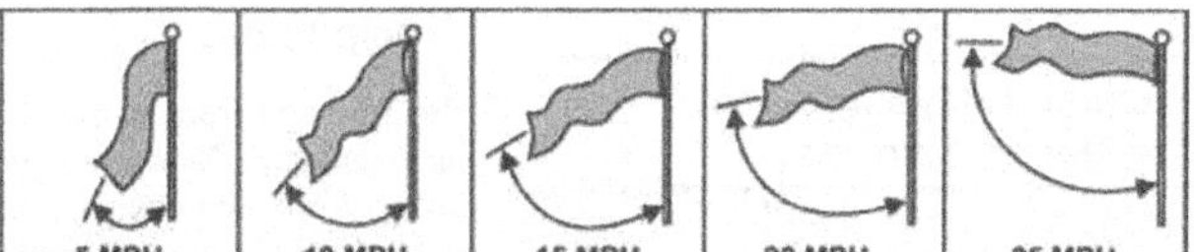

RANGE FLAG ANGLES	5 MPH		10 MPH		15 MPH		20 MPH		25 MPH	
	WIND VALUE		WIND VALUE		WIND VALUE		WIND VALUE		WIND VALUE	
RANGE YARDS	FULL	HALF	FULL	HALF	FULL	HALF	FULL	HALF	FULL	HALF
200	2	1	3	1	5	2	6	3	8	4
300	3	1	6	3	10	5	13	6	16	8
400	4	2	9	4	13	6	18	9	22	11
500	6	3	12	6	18	9	24	12	30	15

Figure 8-5. Windage Click Chart for the Flag Method.

Once the rifle is zeroed, a change in temperature of 20 degrees or more can cause the bullet to strike above or below the point of aim. Therefore, if the temperature changes 20 degrees or more, a Marine should re-zero the rifle.

If the rifle is exposed to below freezing temperatures, it should not be brought immediately into a warm location. Condensation may form on and in the rifle, and it may freeze if re-exposed to the cold. Ice that forms inside the rifle may cause it to malfunction. Ice can form on the rear sight aperture due to condensation, making it impossible to acquire sight alignment.

Precipitation

Freezing rain and other types of precipitation may make the rifle difficult to handle, foul the rifle and cause stoppages, or build up in the barrel or compensator and cause erratic shots. Care should be taken to keep the barrel and muzzle free of water. If the rifle has been submerged, ensure the bore is drained before firing. To drain the bore, pull the charging handle slightly to the rear and hold for a few seconds while the muzzle points down. Once the barrel has been drained, turn the rifle muzzle up to allow the water to drain out of the stock.

8003. Physical and Psychological Effects of Weather on Marines

Wind

Marines can shoot effectively in windy conditions if they apply a few basic techniques and develop the proper mental attitude. The Marine can combat the wind in a number of ways:

- Make subtle changes to the basic shooting positions, such as increasing muscular tension, to reduce movement of the rifle sights.
- Select a more stable firing position.
- Seek support to stabilize the rifle.
- Hold the shot and apply the fundamentals during a lull in the wind.

Temperature

Extreme Heat

In extreme heat, a Marine may experience rapid fatigue. Heat can cause muscle cramps, heat exhaustion, heat stroke, blurred vision, and reduced concentration levels that result in inaccurate shooting. Increased fluid intake and good physical condition can offset the effects of extreme heat. Sweat running into the eyes can cause irritation and make it difficult to see the sights. Extreme heat also can create ground mirages that cause a target to appear indistinct and to drift from side to side. Heat waves or mirages may also distort the target shape or the appearance of the front sight post. A mirage created by the heat of the barrel reduces a Marine's ability to see the sight clearly. To overcome the effects of heat and accurately engage a target, a Marine should maintain a center of mass hold.

Extreme Cold

Extreme cold may affect a Marine's ability to concentrate. If a Marine's hands are numb, he will have difficulty holding a frigid rifle and executing effective trigger control. To protect the hands in a cold environment, a Marine should wear arctic mittens or gloves. To operate the rifle while wearing arctic mittens or gloves, a Marine depresses the trigger guard plunger to open the trigger guard. This allows easier access to the trigger. See figure 8-6. The hasty sling can assist in holding the hand in place on the hand guards so the hand does not slip while wearing mittens.

Precipitation

Precipitation (rain, snow, hail, sleet) can affect target engagement, a Marine's comfort level, and a Marine's ability to concentrate. The amount and type of precipitation may obscure or completely hide the target and it may reduce a Marine's ability to establish an accurate sight picture. Precipitation collecting on rear sight aperture can make it difficult to establish sight alignment and sight picture. Protect sights as much as possible

Figure 8-6. Open Trigger Guard.

during periods of precipitation. It is easy to lose concentration when wet and uncomfortable. Proper dress reduces the effects of precipitation on the Marine.

Light

Light conditions can change the appearance of a target. Light affects each Marine differently. Light can affect range estimation, visual acuity, or the placement of the tip of the front sight on the target. By maintaining a center of mass hold, the effects of light can be reduced.

Bright Light

Bright light conditions exist under a clear blue sky with no fog or haze present to filter the sunlight. Bright light can make a target appear smaller and farther away. As a result, it is easy to overestimate range. Maintaining a center of mass hold, regardless of how indistinct the target appears, ensures the best chances for an effective shot.

Overcast

An overcast condition exists when a solid layer of clouds blocks the sun. The amount of available light changes as the overcast thickens. Overcast conditions can make a target appear larger and closer. As a result, it is easy to underestimate range. During a light overcast, the target appears very distinct and the rifle sights appear very distinct, making it easy to establish sight alignment. As the overcast thickens, it becomes difficult to identify the target from its surroundings.

Haze

Hazy conditions exist when fog, dust, humidity or smoke is present. Hazy conditions can make a target appear indistinct making it difficult to establish sight picture.

Chapter 9. Zeroing

To be combat effective, it is essential for the Marine to know how to zero his rifle. Zeroing is adjusting the sights on the weapon to cause the shots to impact where the Marine aims. This must be done while compensating for the effects of weather and the range to the target. It is critical that Marines can zero their rifles and make the sight adjustments required to engage targets accurately.

See the appropriate technical manual for procedures on boresighting and zeroing with supplemental aiming devices (e.g., laser, night vision devices).

Note

☞ The procedures in this manual are written for right-handed Marines; left-handed Marines should reverse instructions as necessary.

9001. Elements of Zeroing

There are five basic elements involved in zeroing a rifle: line of sight, aiming point, centerline of the bore, trajectory, and range. See figure 9-1.

Line of Sight

The line of sight is a straight line, which begins with the shooter's eye, proceeds through the center of the rear sight aperture, and passes across the tip of the front sight post to a point of aim on a target.

Aiming Point

The aiming point is the precise point where the tip o the front sight post is placed in relationship to target.

Centerline of the Bore

Centerline of the bore is an imaginary straight-line be ginning at the chamber end of the barrel, proceedin out of the muzzle, and continuing indefinitely.

Trajectory

In flight, a bullet does not follow a straight line bu travels in a curve or arc, called trajectory. Trajectory i the path a bullet travels to the target. As the bullet exit the muzzle, it travels on an upward path, intersectin the line of sight (because the sights are above the muz zle). As the bullet travels farther, it begins to drop an intersects the line of sight again.

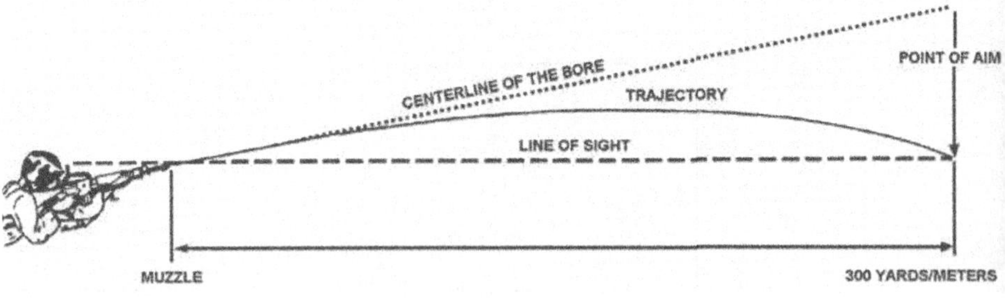

Figure 9-1. Elements of Zeroing.

Range

Range is the KD from the rifle muzzle to the target.

9002. Types of Zeros

Battlesight Zero (BZO)

A BZO is the elevation and windage settings required to place a single shot, or the center of a shot group, in a predesignated location on a target at 300 yards/meters, under ideal weather conditions (i.e., no wind). A BZO is the sight settings placed on your rifle for combat. In combat, your rifle's BZO setting will enable engagement of point targets from 0–300 yards/meters in a no-wind condition.

Zero

A zero is the elevation and windage settings required to place a single shot, or the center of a shot group, in a predesignated location on a target at a specific range, from a specific firing position, under specific weather conditions.

True Zero

A true zero is the elevation and windage settings required to place a single shot, or the center of a shot group, in a predesignated location on a target at a specific range other than 300 yards/meters, from a specific firing position, under ideal weather conditions (i.e., no wind).

9003. M16A2 Sighting System

The sighting system of the M16A2 service rifle consists of a front sight post and two rear sight apertures windage and elevation knob. Scales of the sighting system may be applied accurately to both yard and meter measurements. For example, a rear sight elevation setting of 8/3 may be used for 300 yards/meters.

Front Sight

The front sight post is used to adjust for elevation. The front sight consists of a square, rotating sight post with a four-position, spring-loaded detent (see fig. 9-2). To adjust for elevation, use a pointed instrument (or the tip of a cartridge) to depress the detent and rotate the front sight post (see fig. 9-3). To raise the strike of the bullet, rotate the post clockwise (in the direction of the arrow marked UP) or to the right. To lower the strike of the bullet, rotate the post counter-clockwise (in the opposite direction of the arrow) or to the left.

Figure 9-2. Front Sight.

Figure 9-3. Front Sight Adjustment.

Rear Sight

The rear sight consists of two sight apertures, a windage knob, and an elevation knob. See figure 9-4. The large aperture marked 0-2 is used for target engagement during limited visibility, when a greater field of view is desired, or for engagements of targets closer than 200 yards/meters. The unmarked aperture (small aperture) is used for zeroing and normal firing situations.

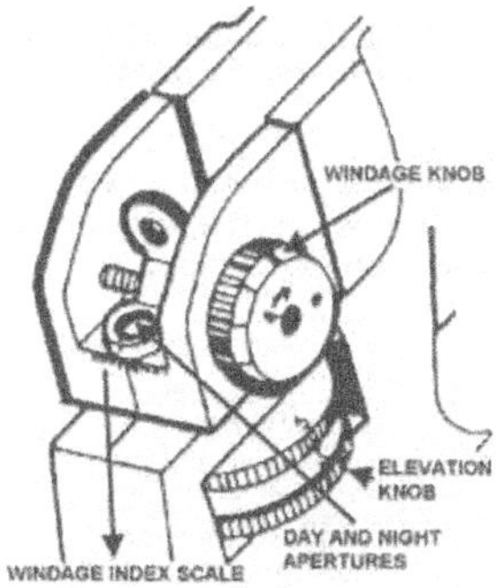

Figure 9-4. Rear Sight.

Elevation Knob

The rear sight elevation knob is used to adjust the sight for a specific range to the target. The elevation knob is indexed as shown in figure 9-5. Each number on the knob represents a distance from the target in 100-yard/meter increments. To adjust for range to the target, rotate the elevation knob so the desired setting is aligned with the index on the left side of the receiver.

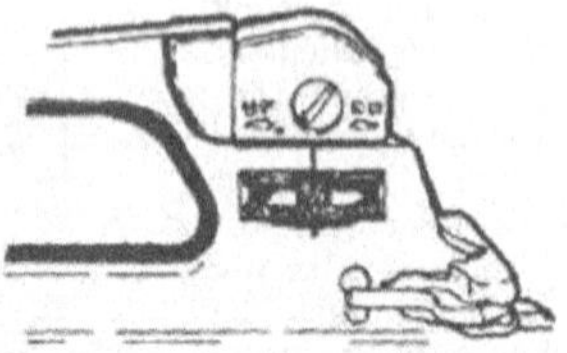

Figure 9-5. Elevation Knob.

If the elevation knob is turned so the number 8/3 aligns with the elevation index line, the 3 indicates 300 yards/meters (see fig. 9-6).

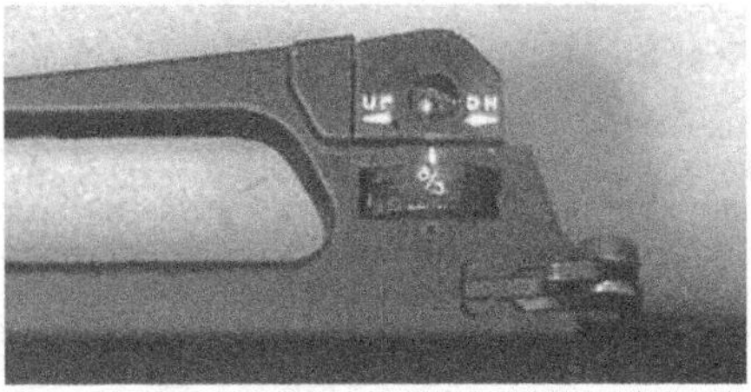

Figure 9-6. Rear Sight Elevation Knob Set for 300 Yards/Meters.

When the rear sight elevation knob is set on 8/3 for 800 yards/meters, there will be a considerable gap (about a quarter inch) between the rear sight housing and the upper receiver (see fig. 9-7).

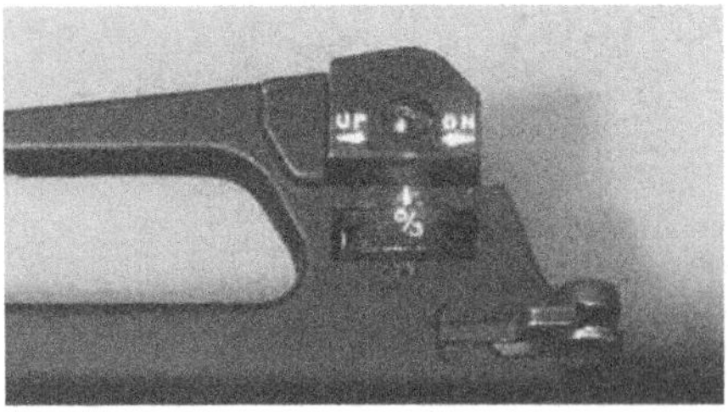

Figure 9-7. Rear Sight Elevation Knob Set for 800 Yards/Meters.

A hasty sight setting is the setting placed on the rear sight elevation knob to engage targets beyond 300 yards/meters. Hasty sight settings for ranges of 400 to 800 yards/meters are applied by rotating the rear sight elevation knob to the number that corresponds with the engagement distance of the enemy. Aligning the number 4, 5, 6 or 7 with the elevation index line places the elevation at 400, 500, 600 or 700 yards/meters, respectively. If a clockwise rotation is continued, the number 8/3 appears for the second time on the elevation index line and indicates an 800 yard/meter elevation.

Windage Knob

The windage knob is used to adjust the strike of the round right or left. The windage knob is marked with an arrow and the letter R that shows the direction the strike of the round is being moved. See figure 9-8.

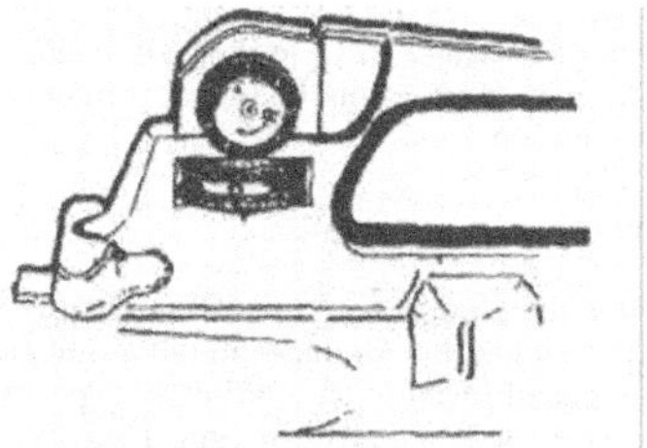

Figure 9-8. Windage Knob.

To move the strike of the round to the right, rotate the windage knob clockwise (in the direction of the arrow marked R).

To move the strike of the round to the left, rotate the windage knob counterclockwise.

9004. Windage and Elevation Rules

Moving the front sight post, elevation knob or windage knob one graduation or notch is referred to as moving one "click" on the sight. The windage and elevation rules define how far the strike of the round will move on the target for each click of front and rear sight elevation or rear sight windage for each 100 yards/meters of range to the target.

Front Sight Elevation Rule

One click of front sight elevation adjustment will move the strike of the round on target approximately 1.25 inches for every 100 yards of range to the target or 3.5 centimeters for every 100 meters of range to the target.

Rear Sight Elevation Rule

One click of rear sight elevation adjustment will move the strike of the round on the target approximately 1 inch for every 100 yards of range to the target or 2.5 centimeters for every 100 meters of range to the target.

Windage Rule

One click of windage adjustment will move the strike of the round on the target approximately 0.5 inch for every 100 yards of range to the target or 1.25 centimeters for every 100 meters of range to the target.

9005. Initial Sight Settings

Initial sight settings are those settings that serve as the starting point for initial zeroing from which all sight adjustments are made. (If the Marine already has a BZO established on his rifle, he may begin the zeroing process by using the previously established BZO sight settings.) To set the sights to initial sight settings:

Front Sight Post

To set the front sight post to initial sight setting, depress the front sight detent and rotate the front sight post until the base of the front sight post is flush with the front sight housing.

Rear Sight Elevation Knob

To set the elevation knob at the initial sight setting, perform the following:

- Rotate the rear sight elevation knob counterclockwise until the moveable rear sight housing is bottomed out on the upper receiver. See figure 9-9.

Figure 9-9. Bottoming Out Elevation Knob.

Note

Once bottomed out, the rear sight elevation knob should be three clicks counterclockwise from 8/3. If the sight fails to move three clicks counterclockwise from 8/3, it must be adjusted by a qualified armorer.

- Rotate rear sight elevation knob clockwise until the number 8/3 aligns with the index mark located on the left side of the upper receiver. See figure 9-10.

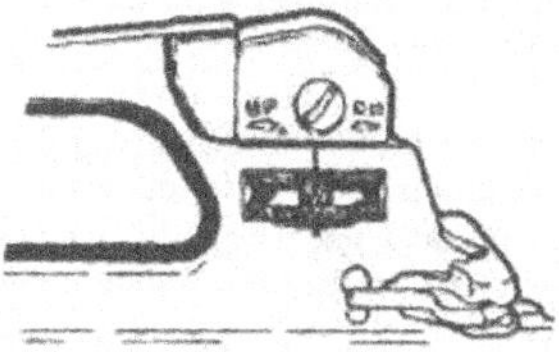

Figure 9-10. Elevation Knob Set at 8/3.

Windage Knob

To set the windage knob to initial sight setting, rotate windage knob until the index line located on the top of the large rear sight aperture aligns with the centering on the windage index scale located on the moveable base of the rear sight assembly. See figure 9-11.

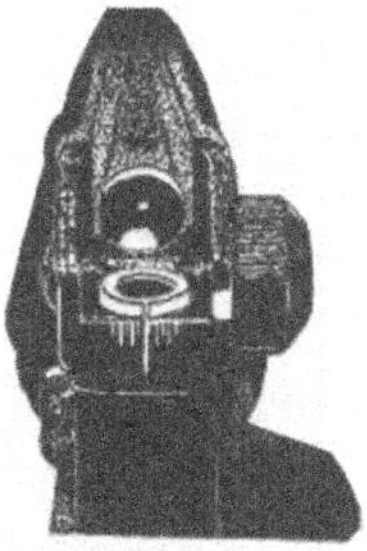

Figure 9-11. Aligning Index Line.

9006. Zeroing Process

During the zeroing process, all elevation adjustments are made on the front sight post. Once a BZO is established, the front sight post should never be moved, except when rezeroing the rifle. (The rear sight elevation knob is used for dialing in the range to the target.) Zeroing is conducted at a range of 300 yards/meters. To prepare a rifle for zeroing, the rifle sights must be adjusted to the initial sight settings as outlined in paragraph 9005. Perform the following steps to zero the rifle:

- Fire a 3-shot group.
- Triangulate the shot group to find the center. See figure 9-12.
- Determine the vertical distance in inches from the center of the shot group to the center of the target.
- Make elevation adjustments on the front sight post to move the center of the shot group to the center of the target.
- Determine the horizontal distance from the center of the shot group to the center of the target.
- Make lateral adjustments on the windage knob to move the center of the shot group to the center of the target.
- Repeat preceding steps until shot group is centered.
- Fire a 4-round shot group to confirm sight setting.

Once the sight setting is confirmed, determine the value and direction of the wind and remove the number of clicks added to the windage knob (if necessary) to compensate for current wind conditions. This becomes the BZO setting for the rifle.

9007. Battlesight Zero

Zeroing is conducted at a range of 300 yards/meters. If a 300-yard/-meter range is not available, a field expedient BZO can be established at a reduced range of 36 yards/30 meters. When a rifle is zeroed for 300 yards/meters, the bullet crosses the line of sight twice. It first

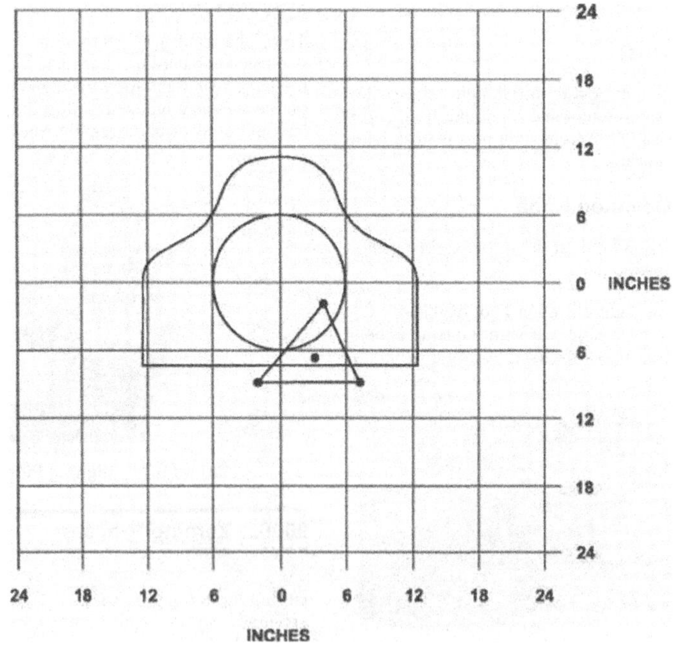

Figure 9-12. Triangulating Shot Group.

crosses the line of sight on its upward path of trajectory at 36 yards/30 meters, and again farther down range at 300 yards/meters (see fig. 9-13). Therefore a rifle's BZO may be established at a distance of 36 yards/30 meters and the same BZO will be effective at 300 yards/meters.

To establish a field expedient BZO at 36 yards or 30 meters when a 300-yard/-meter range is not available, a Marine performs the same steps as the zeroing process outlined in paragraph 9006. However, since wind does not affect the round at 36 yards/30 meters, windage is not added nor is it removed from the windage knob after confirming the BZO. To be accurate, the target must be placed exactly 36 yards (or 30 meters) from the muzzle of the rifle.

9008. Factors Causing a BZO to be Reconfirmed

Marines are responsible for maintaining a BZO on their rifles at all times. Many factors influence the BZO of a rifle. Atmospheric conditions, humidity, and temperature can cause BZOs to change on a daily basis. If operating in a combat environment, Marines should confirm their BZO as often as possible. To confirm a BZO, a Marine may begin the zeroing process by using the previously established BZO sight settings rather than placing the sights at initial sight setting. The following factors cause a BZO to be reconfirmed.

Maintenance

It is possible for the BZO to change if ordnance personnel perform maintenance on a rifle. If maintenance was performed, it is critical that the rifle be rezeroed as soon as possible.

Temperature

An extreme change in temperature (i.e., 20 degrees or more) will cause the elevation BZO to change. Changes in temperature cause chamber pressure to increase when hot and decrease when cold. This causes shots to impact the target high in hot temperatures and low in cold temperatures.

Climate

Changing climates (i.e., moving from a dry climate to a tropical climate) can mean changes in air density, moisture content, temperature or barometric pressure. Any of these elements can affect the rifle's BZO.

Ammunition

Inconsistencies in the production of ammunition lots can change a rifle's BZO.

Ground Elevation

Drastic changes in ground elevation can create changes in air density, moisture content, temperature or barometric pressure. Any of these elements can affect the rifle's BZO.

Uniform

If Marines zero their rifles in utility uniform and fire in full battle gear, their BZOs will change. The wearing of full battle gear changes eye relief, placement of the rifle in the shoulder pocket, and the way the rifle is supported on the handguard. Marines must establish their BZOs while wearing the uniform and equipment they will be wearing while engaging targets.

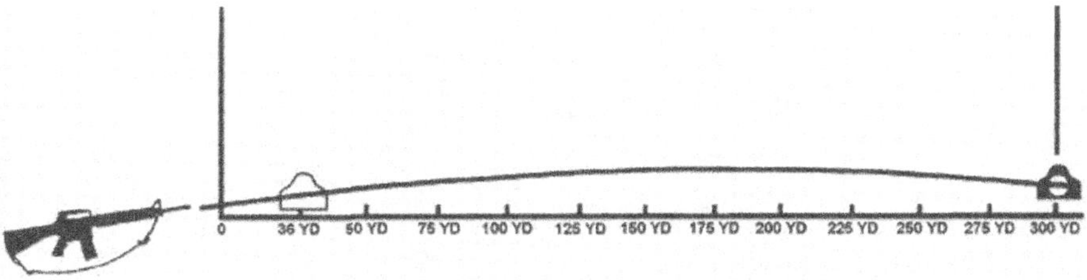

Figure 9-13. Bullet Crossing the Line of Sight Twice.

9009. Factors Affecting the Accuracy of a BZO

Anything the Marine changes from shot to shot affects the accuracy of his BZO. It further affects the accuracy of shot placement. The following factors, when applied inconsistently, diminish the accuracy of a BZO:

- Any of the seven factors (left hand, rifle butt in th pocket of the shoulder, grip of the right hand, righ elbow, stock weld, breathing, and muscular relax ation/tension).
- Stability of hold.
- Sling tension.
- Trigger control.
- Sight picture.

Chapter 10. Engagement Techniques

A Marine must maintain the ability to react instinctively in a combat environment—day or night. He must possess a combat mindset that eliminates hesitation, fear, or uncertainty of action. A Marine's combat mindset allows him to engage the enemy rapidly and focus on the actions required to fire well-aimed shots. He must remember that speed alone does not equate to effective target engagement. He should fire only as fast as he can fire accurately. He should never exceed his physical capability to engage a target effectively. To be effective in combat, a Marine must train to perfect the physical skills of target engagement (such as presenting the weapon and assuming a shooting position) until they become instinctive. A Marine must also employ effective engagement techniques that enable acquisition and engagement of a variety of targets in diverse combat conditions.

10001. Target Detection

To be proficient, a Marine rifleman must be able to detect targets, determine the range to targets, and accurately engage the targets. There are many variables affecting a Marine's ability to detect and determine the range to combat targets. Enemy targets on the battlefield may be single or multiple, stationary or moving, or completely hidden from view. Success in locating an enemy target will depend upon the observer's position, his skill in searching an area, and his ability to recognize target indicators.

Target Indicators

Most combat targets are detected at close range by smoke, flash, dust, noise or movement, and are usually seen only momentarily. Target indicators are anything that reveals an individual's position to the enemy. These indicators are grouped into three general areas: movement, sound, and improper camouflage.

Movement

The human eye is attracted to movement, especially sudden movement. The Marine need not be looking directly at an object to notice movement. The degree of difficulty in locating moving targets depends primarily on the speed of movement. A slowly moving target will be harder to detect than one with quick, jerky movements.

Sound

Sound can also be used to detect an enemy position. Sound may be made by movement, rattling equipment, or talking. Sound provides only a general location of the enemy, making it difficult to pinpoint a target by sound alone. However, sound can alert the Marine to the presence of a target and increase his probability of locating it through other indicators.

Improper Camouflage

There are three indicators caused by improper camouflage: shine, outline, and contrast with the background. Most targets on the battlefield are detected due to improper camouflage. However, many times an observation post or enemy firing position will blend almost perfectly with the natural background. Only through extremely careful, detailed searching will these positions be revealed.

Shine. Shine is created from reflective objects such as metal or glass. It may also come from pools of water and even the natural oils from the skin. Shine acts as a beacon to the target's position.

Outline. Most enemy soldiers will camouflage themselves, their equipment, and their positions. The outline of objects such as the body, head and shoulders, weapons, and web gear are recognizable even from a distance. The human eye will often pick up a recognizable shape and concentrate on it even if the object cannot be identified immediately. The reliability of this indicator depends upon visibility and the experience of the observer.

Contrast With the Background. Indicators in this category include objects that stand out against (contrast with) a background because of differences in color, surface, and shape. For instance, a target wearing a dark uniform would be clearly visible in an area of snow or sand. Geometric shapes, such as helmets or rifle barrels, can be easy to detect in a wooded area. Fresh soil around a fighting hole contrasts with the otherwise unbroken ground surface. While observing an area, take note of anything that looks out of place or unusual and study it in more detail. This will greatly increase your chances of spotting a hidden enemy.

Identifying Target Location

Observation Position

A good position is one that offers maximum visibility of the area while affording cover and concealment. The optimal observation position should allow the Marine to scan all the areas of observation and offer enough concealment to prevent his position from being detected.

The Marine should avoid positions that are obvious or stand out, such as a lone tree in a field or a pile of rocks on a hill. These positions may be ideal points for easy observation, but they will also make it easier for the enemy to locate the Marine.

Methods for Searching an Area

In searching an area, the Marine will be looking for target indicators. There are two techniques for searching an area: the hasty search and the detailed search.

Hasty Search. When a Marine moves into a new area, he must quickly check for enemy activity that may pose an immediate danger. This search is known as the hasty search and should take about 30 seconds, depending on the terrain.

Quickly glance at various points throughout the area rather than sweeping the eyes across the terrain in one continuous movement. The Marine should search the area nearest him first since it poses the greatest potential for danger.

This method of search is effective because it takes advantage of peripheral vision. Peripheral vision enables the detection of any movement in a wide area around the object being observed. For this technique to be effective, the eyes must be focused briefly on specific points (i.e., areas that may provide cover or concealment for the enemy).

Detailed Search. A detailed search is a systematic examination of a specific target indicator or of the entire observation area. A detailed search should be conducted immediately on target indicators located during the hasty search. The detailed search should be made from top to bottom or side to side, observing the entire object in exact detail. If multiple indicators were observed during the hasty search, the detailed search should begin with the indicator that appears to pose the greatest threat.

After a thorough search of target indicators, or if no indicators were located during the hasty search, a detailed search should be made of the entire observation area. The 50-meter overlapping strip method is normally used.

Normally, the area nearest the observer offers the greatest potential danger and should be searched first. Begin the search at one flank, systematically searching the terrain at the front in 180-degree arcs, searching everything in exacting detail, 50 meters in depth. See figure 10-1.

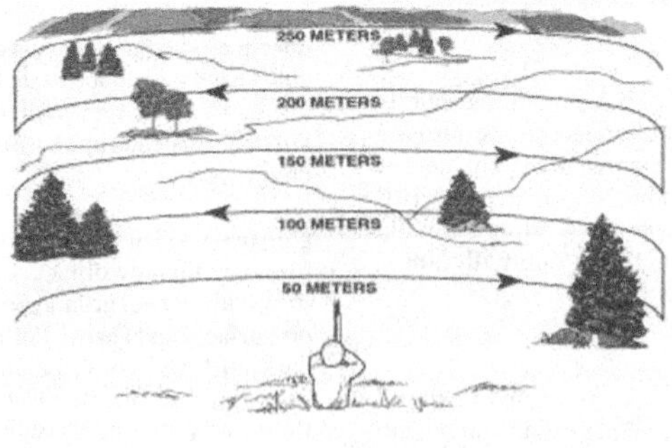

Figure 10-1. Searching the Terrain in Overlapping Strips.

After reaching the opposite flank, systematically cover the area between 40 and 90 meters from your position. The second search of the terrain includes about 10 meters of the area examined during the first search. This technique ensures complete coverage of the area.

Continue the overlapping strip search method for as far as you can see.

Maintaining Observation

Method

The combat situation will dictate the method of maintaining observation of an area. Generally, the observation method will include a combination of hasty and detailed searches.

Sequence of Observation

Observation is often conducted by a two-man team. One team member should constantly observe the entire area using the hasty search technique and the other team member should conduct a detailed overlapping strip search. If you are observing as an individual, devise a plan to ensure that the area of observation is completely covered. When entering a new area, immediately conduct a hasty search. Since a hasty search may fail to detect some indicators, periodically conduct a detailed search of the area. A detailed search should also be conducted any time your attention has been diverted from the search area.

Remembering Target Location

Most targets are seen only briefly and most areas contain multiple targets. Once you have located a target indicator, you will need to remember its location to engage it successfully. To help remember the location of a target, select a known feature and use it as a reference point to determine the distance and general direction to the target.

10002. Range Estimation

To engage targets at unknown distances, a Marine must determine the distance from his location to a known point. This is known as range estimation. The ability to determine range is a skill that must be developed if a Marine is to successfully engage targets at unknown distances. Precise range estimation enhances accuracy, enhances the chance of survival, and determines if a target can be effectively engaged using the rifle's existing BZO or if a new sight setting or point of aim is required.

Range Estimation Methods

Unit of Measure Method

To use this method, a Marine visualizes a distance of 100 meters on the ground, and then estimates how many of these units will fit between him and the target. This determines the total distance to the target. See figure 10-2.

Figure 10-2. Unit of Measure Method.

The greatest limitation of this method is that its accuracy is related to the amount of visible terrain. For example, if a target appears at a range of 500 meters or more and only a portion of the ground between a Marine and the target can be seen, it becomes difficult to use the unit of measure method to estimate range accurately. A Marine must practice this method frequently to be proficient. Whenever possible, a Marine should select an object, estimate the range, and then verify the actual range by either pacing or using another accurate measurement.

Rifle Front Sight Post Method

The area of the target covered by the rifle's front sight post can be used to estimate range to a target. A Marine notes the appearance of the front sight post on a known-distance target. A Marine then uses this as a guide to determine range over an unknown distance. Because the apparent size of the target changes as the distance to the target changes, the amount of the target covered by the front sight post varies based on the range. In addition, a Marine's eye relief and perception of the front sight post affect the amount of the target that is visible. To use this method, a Marine must apply the following guidelines:

- The front sight post covers the width of a man's chest or body at approximately 300 meters.
- If the target is less than the width of the front sight post, the target is in excess of 300 meters and the rifle's BZO cannot be used effectively.
- If the target is wider than the front sight post, the target is less than 300 meters and can be engaged point of aim/point of impact using the rifle's BZO.

See figure 10-3.

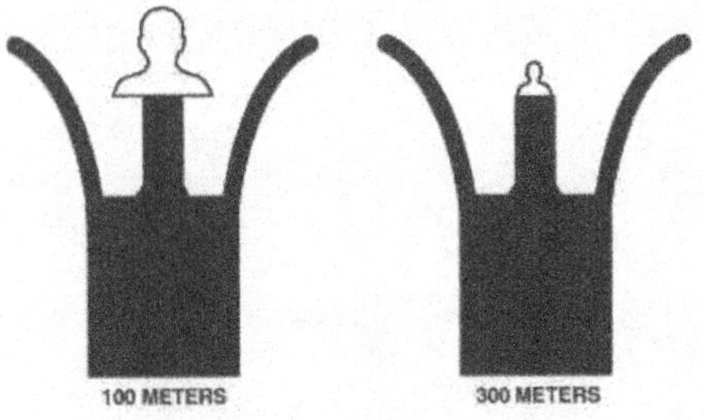

Figure 10-3. Front Sight Post Method.

Visible Detail Method

The amount of detail seen at various ranges can provide a Marine with an estimate of the target's distance. To use this method, a Marine must be familiar with the size and various details of personnel and equipment at known distances. Visibility (such as weather, smoke or darkness) limits the effectiveness of this method. A Marine should observe a man while he is standing, kneeling, and in the prone position at known ranges of 100 to 500 meters. He should note the man's size, characteristics/size of his uniform and equipment, and any other pertinent details. The Marine then uses this as a guide to determine range over an unknown distance. A Marine also should study the appearance of other familiar objects such as rifles and vehicles. To use this method, a Marine applies the following general guidelines:

- At 100 meters, the target is clearly observed in detail and facial features are distinguished
- At 200 meters, the target is clearly observed. There is a loss of facial detail. The color of the skin and equipment are still identifiable.
- At 300 meters, the target has a clear body outline, face color usually remains accurate, but remaining details are blurred.
- At 400 meters, the body outline is clear, but remaining detail is blurred.
- At 500 meters, the body shape begins to taper at the ends, and the head becomes indistinct from the shoulders.
- At 600 meters, the body appears wedge-shaped and headless.

Bracketing Method

This method of range estimation estimates the shortest possible distance and the greatest possible distance to the target. For example, a Marine estimates that a target may be as close as 300 meters but it could be as far away as 500 meters. The estimated distances are averaged to determine the estimated range to the target. For example, the average of 300 meters and 500 meters is 400 meters.

Halving Method

To use the halving method, a Marine estimates the distance halfway between him and the target, then doubles that distance to get the total distance to the target. A Marine must take care when judging the distance to the halfway point; any error made in judging the halfway distance is doubled when estimating the total distance.

Combination Method

The methods previously discussed require optimal conditions with regard to the target, terrain, and visibility in order to obtain an accurate range estimation. A Marine should estimate the range using two methods and then compare the estimates, or two Marines can compare their estimates. The average of the two estimates should be close to the actual range to target.

Factors Affecting Range Estimation

The following specific factors will affect the accuracy of estimation. A Marine must be aware of these factors and attempt to compensate for their effects.

Nature of the Target

- An object with a regular outline such as a steel helmet, rifle, or vehicle on a clear day will appear to be closer than one with an irregular outline such as a camouflaged object.

- A target that contrasts with its background will appear to be closer than a target that blends in with its background.
- A partially exposed object will appear to be farther away than it is.
- A target will appear to be farther away if the target is smaller than the objects surrounding it.

Nature of the Terrain

- Terrain that slopes upward gives the illusion of shorter distance.
- Terrain that slopes downward gives the illusion of greater distance.
- Terrain with dead space makes the target appear to be closer.
- Smooth terrain such as sand, water, or snow gives the illusion of greater distance.

Limited Visibility

The position of the light source significantly affects a Marine's ability to estimate range. Other factors that affect range estimation include smoke, fog, rain, angled light, or anything that obscures the battlefield. Generally, when the sun is bright, a target appears further away. When the sky is overcast, a target generally appears closer. Target contrast is another factor to consider when estimating range as well as obstacles located between the shooter and the target. If there is contrast (e.g., color variation) between the target and the background, the target will appear closer. If there is little or no contrast between the target and the background, the target will appear farther away. If there is an object between the target and the shooter that will distract the shooter (e.g., a tree), the target will appear farther away.

10003. Offset Aiming

The conditions of rifle fire in combat may not permit mechanical adjustments of the sights. To engage a target during combat, a Marine may be required to aim his rifle at a point on the target other than center mass. This is known as offset aiming. Offset aiming is used to compensate for the distance and size of the target, wind, and speed and angle of a moving target. There are two primary techniques for offset aiming: point of aim technique and known strike of the round.

Point of Aim Technique

The point of aim technique is the shifting of the point of aim (sight picture) to a predetermined location on or off the target to compensate for a known condition (i.e., wind, distance, and movement). Each predetermined location is known as a point of aim. Figure 10-4 illustrates points of aim for elevation.

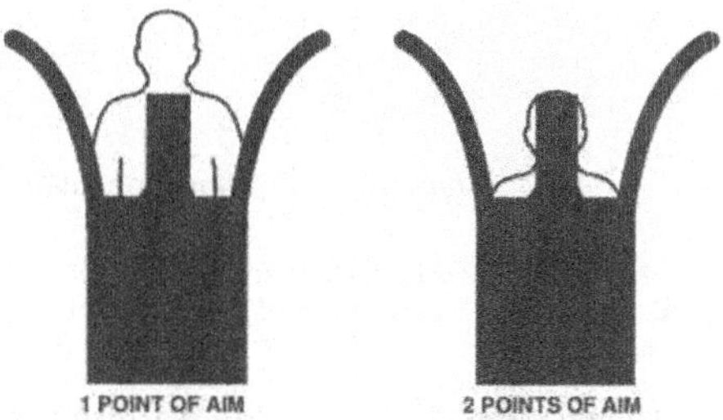

Figure 10-4. Points of Aim for Elevation.

Elevation

Predetermined points of aim sector the target horizontally. The tip of the front sight post held at shoulder level is considered one point of aim; the tip of the front sight post held at the top of the target's head is considered two points of aim. A Marine uses these points of aim to compensate for the elevation required to engage a target beyond the BZO capability of the rifle or to engage a small target (e.g., head shot) inside the BZO of the weapon.

Beyond the BZO. To use the point of aim technique to engage a target beyond the BZO of the rifle, a Marine must apply the following guidelines:

- When range to the target is estimated to be beyond 300 meters out to 400 meters, hold one point of aim.
- When the range to the target is estimated to be beyond 400 meters out to 500 meters, hold two points of aim.

Note

It is better to apply a hasty sight setting at ranges beyond the rifle's BZO. Points of aim are only guidelines at these distances because the front sight will mask the target when the front sight is held above center mass, making it difficult to acquire sight picture.

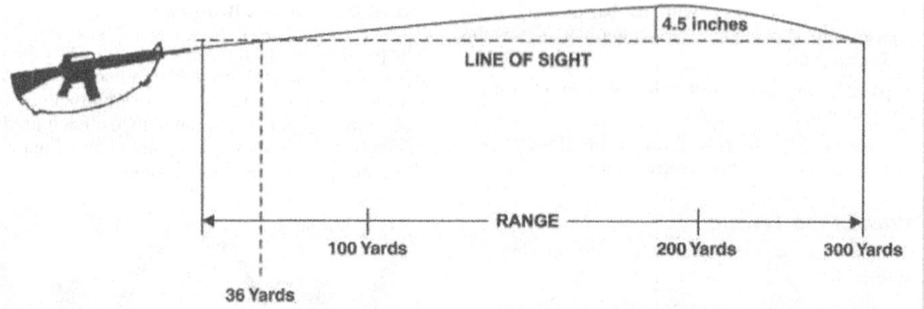

Figure 10-5. Trajectory and Point of Aim/Point of Impact for 300 Yard BZO.

Inside the BZO. If the rifle is properly zeroed for 300 yards/meters, the trajectory (path of the bullet) will rise approximately 4.5 inches (11 centimeters) above the line of sight at a distance of approximately 175 yards (160 meters). At other distances, the strike of the bullet will be less than 4.5 inches above the point of aim. Only at 36 yards/30 meters and 300 yards/meters does the point of impact coincide with the point of aim. If only a portion of the target is visible (e.g., the head of an enemy soldier), the trajectory of the bullet may have to be taken into consideration when firing at a distance less than 300 yards/meters. If a Marine does not consider trajectory, he may shoot over the top of the target if the target is small and at a distance less than 300 yards/meters. See figure 10-5.

Windage

Predetermined points of aim sector the target vertically (see fig. 10-6). The tip of the front sight post centered on the edge of the target into the wind is considered one point of aim; the trailing edge of the front sight post held on the edge of the target into the wind is considered two points of aim. The same units of measure are applied off the target for holds of additional points of aim. These points of aim are used to compensate for wind affecting the strike of the round and when there is no time to adjust the rifle's sights, or when a lead is required to engage a moving target (points of aim for moving targets are discussed in paragraph 10007). Table 10-1 provides points of aim for full value winds.

Known Strike of the Round

This offset aiming technique shifts the aiming point (sight picture) to compensate for rounds that strike off target center. The known strike of the round method is used if the strike of the round is known. To engage a target using this method, a Marine aims an equal distance from center mass opposite the known strike of the round. For example, if the round strikes high and left, a Marine aims an equal and opposite distance low and right.

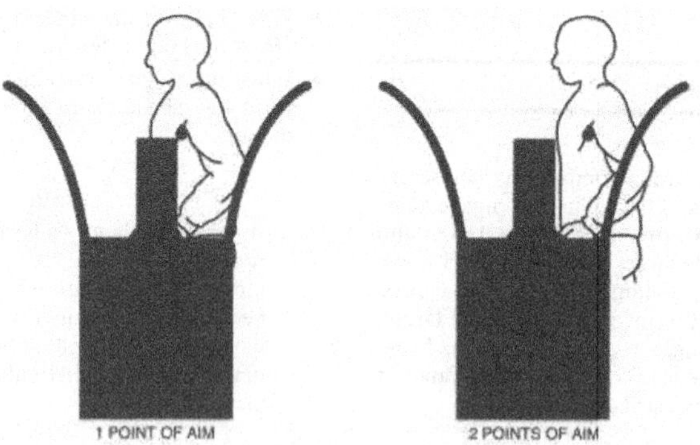

Figure 10-6. Points of Aim.

Table 10-1. Points of Aim for Full Value Winds.

Wind	Distance (yards/meters)	Points of Aim
Light	0-300	0
Medium	200-300	1
Strong	0-200	1
Strong	200-300	2

10004. Techniques of Fire

The size and distance to the target should dictate the technique of fire.

Two-Shot Technique

In combat, an effective technique for eliminating a threat is to rapidly fire more than one shot on the target. Two shots fired in rapid succession will increase the trauma (i.e., shock, blood loss) on the target, increasing the Marine's chances of quickly eliminating the threat. Firing two shots enables the Marine to break out of the tunnel vision often associated with firing in combat and then assess the situation to determine follow-on action. To execute the two-shot technique, the Marine acquires sight picture for each shot fired (each pull of the trigger).

Single Shot Technique

If the target is at a long range or if the target is small (i.e., partially exposed), it can best be engaged with a single, precision shot. A Marine's stability of hold and sight alignment are more critical to accurate engagement of long-range or small targets. To engage a target with the single shot technique, the Marine must slow down the application of the fundamentals and place one well-aimed shot on target.

Sustained Rate of Fire

An effective method for delivering suppressive fire is to fire at the sustained rate of 12 to 15 rounds per minute. Management of recoil is critical to bring the sights back on target after the shot is fired.

Three-Round Burst Technique

When set on burst, the design of the rifle permits three shots to be fired from a single trigger pull. The rounds fire as fast as the weapon will function and cause the muzzle to climb during recoil. The ability to manage recoil is extremely important when firing the rifle on burst. To achieve the desired effect (i.e., 3 rounds on target), the Marine must control the jump angle of the weapon to maintain the sights on target. At short ranges (i.e., 25 meters or less), firing on three-round burst can be an effective technique to place rounds on a man-sized target quickly to increase trauma on the target. To execute the three-round burst technique, the Marine places the selector lever on burst, aims center mass, and acquires sight picture once for the single trigger pull.

10005. Engaging Immediate Threat Targets

Immediate threat target engagement is characterized by short-range engagement (i.e., less than 50 meters) with little or no warning that requires an immediate response to engage an enemy. This type of engagement is likely in close terrain (e.g., urban, jungle). If this type of engagement is likely, the large rear sight aperture (0-2) could be placed up to provide a wider field of view and detection of targets. Marksmanship skills include quick presentation and compression of the fundamentals (i.e., quick acquisition of sight picture, uninterrupted trigger control). At close ranges, perfect sight alignment is not as critical to accuracy on target. However, the front sight post must be in the rear sight aperture; proper sight alignment is always the goal.

10006. Engaging Multiple Targets

When engaging multiple targets, a Marine must prioritize each target and carefully plan his shots to ensure successful target engagement. Mental preparedness and the ability to make split-second decisions are the key to successful engagement of multiple targets. The proper mindset allows a Marine to react instinctively and to control the pace of the battle rather than just reacting to the threat.

After the first target is engaged, a Marine must immediately engage the next target and continue to engage targets until they are eliminated. While engaging multiple targets, a Marine must be aware of his surroundings and not fixate on just one target. He must rapidly prioritize the targets, establish an engagement sequence, and engage the targets. A Marine also must maintain constant awareness and continuously search the terrain for additional targets.

Prioritizing Targets

The combat situation will usually dictate the order of multiple target engagement. Target priority is based on factors such as proximity, threat, and opportunity; no two situations will be the same. The principal method is to determine the level of threat for each target so all may be engaged in succession from the most threatening to the least threatening. The target that poses the greatest threat (e.g., closest, greatest firepower) should be engaged first. Prioritizing targets is an ongoing process. Changes in threat level, proximity, or the target itself may cause a Marine to revise his priorities. Therefore, a Marine must remain alert to changes in a target's threat level and proximity and other target opportunities as the battle progresses.

Technique of Engagement

To engage multiple targets, the Marine performs the following steps:

- The first target with two rounds.
- The recoil of the rifle can be used to direct the recovery of the weapon on to the next target. As the weapon is coming down in its recovery, the Marine physically brings the sights onto the desired target. Pressure is maintained on the trigger throughout recovery and trigger control is applied at a rate consistent with the Marine's ability to establish sight picture on the desired target.
- When possible, such as when all targets are of equal threat, the Marine should engage targets in a direction that maximizes his ability to support and control the weapon.
- The preceding steps are repeated until all targets have been engaged.

Firing Position

The selection and effective use of a firing position are critical to the successful engagement of multiple targets. A Marine should make a quick observation of the terrain and select a firing position that provides good cover and concealment, as well as the flexibility to engage multiple targets. If enemy targets are widely dispersed, the selected position must provide the Marine with flexibility of movement. The more restrictive the firing position, the longer it will take a Marine to eliminate multiple targets.

Prone

The prone position limits left and right lateral movement and is, therefore, not recommended for engaging short-range dispersed targets. Because the elbows are firmly placed on the ground in the prone position, upper body movement is restricted.

Sitting

Like the prone position, the sitting position allows limited lateral movement. This makes engagement of widely-dispersed multiple targets difficult. To ease engagement, the forward arm can be moved by pivoting on the elbow, but this movement disturbs the stability of the position.

Kneeling

The kneeling position provides a wider, lateral range of motion since only one elbow is used for support. A Marine moves from one target to another by rotating at the waist to move the forward arm in the direction of the target, either right or left.

Standing

The standing position allows maximum lateral movement. Multiple targets are engaged by rotating the upper body to a position where the sights can be aligned on the desired target. If severe or radical adjustments are required to engage widely dispersed targets, a Marine moves his feet to establish a new position rather than give up maximum stability of the rifle. This avoids poorly placed shots that can result from an unstable position.

10007. Engaging Moving Targets

In combat, it is unlikely that a target will remain stationary. The enemy will move quickly from cover to cover, exposing himself for the shortest possible time. Therefore, a Marine must quickly engage a moving target before it disappears.

Types of Moving Targets

There are two types of moving targets: steady moving targets and stop and go targets. Steady moving targets move in a consistent manner and remain in a Marine's field of vision. A walking or running man is an example of a steady moving target. A stop and go target appears and disappears during its movement. A stop and go target will present itself for only a short time before it reestablishes cover. An enemy moving from cover to cover is an example of a stop and go target. This target is most vulnerable to fire at the beginning and end of its movement to new cover because the target must gain momentum to exit its existing cover and then slow down to occupy a new position.

Leads

When a shot is fired at a moving target, the target continues to move during the time the bullet is in flight. Therefore, a Marine must aim in front of the target; otherwise, the shot will fall behind the target. This is called leading a target. Lead is the distance in advance of the target that the rifle sights are placed to accurately engage the target when it is moving.

Amount of Lead Required

Factors that affect the amount of lead are the target's range, speed, and angle of movement.

Range. Lead is determined by the rifle's distance to the target. When a shot is fired at a moving target, the target continues to move during the time the bullet is in flight. This time of flight could allow a target to move out of the bullet's path if the round was fired directly at the target. Time of flight increases as range to the target increases. Therefore, the lead must be increased as the distance to the target increases.

Speed. If a man is running, a greater lead is required because the man will move a greater distance than a walking man will while the bullet is in flight.

Angle of Movement. The angle of movement across the line of sight relative to the flight of the bullet determines the type (amount) of lead.

Three Types of Leads: Full, Half, No

- **Full Lead**. The target is moving straight across a Marine's line of sight with only one arm and half the body visible. This target requires a full lead because it will move the greatest distance across a Marine's line of sight during the flight of the bullet.
- **Half Lead**. The target is moving obliquely across a Marine's line of sight (at a 45-degree angle). One arm and over half the back or chest are visible. This target requires half of a full lead because it will move half as far as a target moving directly across a Marine's line of sight during the flight of the bullet.
- **No Lead**. The target is moving directly toward or away from a Marine and presents a full view of both arms and the entire back or chest. No lead is required. A Marine engages this target as if it were a stationary target because it is not moving across his line of sight.

Point of Aim Technique

Predetermined points of aim sector the target vertically. The tip of the front sight post centered on the leading edge of the target is considered one point of aim; the trailing edge of the front sight post held on the leading edge of the target is considered two points of aim. The same units of measure are applied off the target for holds of additional points of aim. These points of aim are used to compensate when a lead is required to engage a moving target. The following guidelines apply if a Marine uses the point of aim technique to establish a lead for a moving target at various ranges and speeds (see fig. 10-7, on page 10-10). These guidelines do not consider wind or other effects of weather. Body width in these examples is considered to be 12 inches (30 centimeters) (side view of the target).

For a slow walking target (approximately 2 to 2.5 mph/3.2 to 4 kph) moving directly across the line of sight (full lead)—

- At a range of 200 meters or less, no lead is required.
- At a range of 300 meters, hold one point of aim in the direction the target is moving.

For a fast walking target (approximately 4 mph/6.4 kph) moving directly across the line of sight (full lead)—

- At a range of 200 meters or less, hold one point of aim in the direction the target is moving.
- At a range of 300 meters, hold two points of aim in the direction the target is moving.

For a target running (approximately 6 mph/9.7 kph) directly across the line of sight (full lead)—

- At a range of 100 meters or less, hold one point of aim in the direction the target is moving.

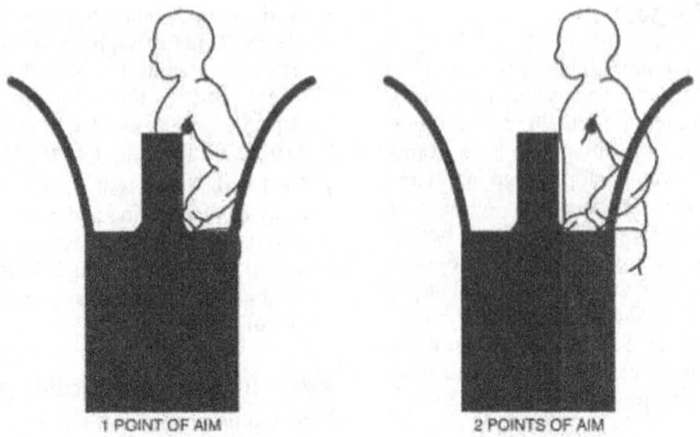

Figure 10-7. Points of Aim.

- At a range of 200 meters, hold two points of aim in the direction the target is moving.

For a target moving at a 45-degree angle (an oblique target) across the line of sight, the lead is one half of the lead that is required for a target moving directly across the line of sight.

Engagement Methods

Moving targets are the most difficult targets to engage. However, they can be engaged successfully by using the tracking or the ambush method.

The Tracking Method

The tracking method is used for a target that is moving at a steady pace over a well-determined route. If a Marine uses the tracking method, he tracks the target with the rifle's front sight post while maintaining sight alignment and a point of aim on or ahead of (leading) the target until the shot is fired. When establishing a lead on a moving target, rifle sights will not be centered on the target and instead will be held on a lead in front of the target. See figure 10-8. A Marine performs the following steps to execute the tracking methods:

- Present the rifle to the target.
- Swing rifle muzzle through the target (from the rear of the target to the front) to the desired lead (point of aim). The point of aim may be on the target or some point in front of the target depending upon the target's range, speed, and angle of movement.
- Track and maintain focus on the front sight post while acquiring the desired sight picture. It may be necessary to shift the focus between the front sight post and the target while acquiring sight picture, but the focus must be on the tip of the front sight post when the shot is fired.
- Engage the target once sight picture is acquired while maintaining the proper lead.
- Follow-through so the lead is maintained as the bullet exits the muzzle.
- Continue to track in case a second shot needs to be fired on the target.

The Ambush Method

The ambush method is used when it is difficult to track the target with the rifle, as in the prone, sitting, or any supported position. The lead required to effectively engaging the target determines the engagement point. With the sights settled, the target moves into the predetermined engagement point and creates the desired sight picture. See figure 10-9.

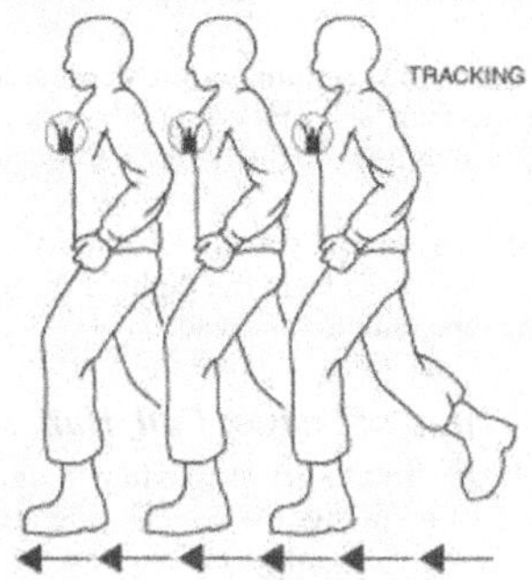

Figure 10-8. Tracking Method.

The trigger is pulled simultaneously with the establishment of sight picture. To execute the ambush method, a Marine performs the following steps:

- Select an aiming point ahead of the target.
- Obtain sight alignment on the aiming point.
- Hold sight alignment until the target moves into vision and the desired sight picture is established.
- Engage the target once sight picture is acquired.
- Follow-through so the rifle sights are not disturbed as the bullet exits the muzzle.

A variation of the ambush method can be used when engaging a stop and go target. A Marine should look for a pattern of exposure (e.g., every 15 seconds). Once a pattern is determined, a Marine establishes a lead by aiming at a point in front of the area in which the target is expected to appear, then he fires the shot at the moment the target appears.

Marksmanship Fundamentals

Engaging moving targets requires concentration and adherence to the fundamentals of marksmanship. The following modifications to the fundamentals of marksmanship are critical to engagement of moving targets.

Sight Picture

Typically, sight picture is the target's center of mass. If a Marine engages a moving target, he bases his sight picture on the target's range, speed, and angle of movement, i.e., sight alignment may be established on a point of aim in front of the target.

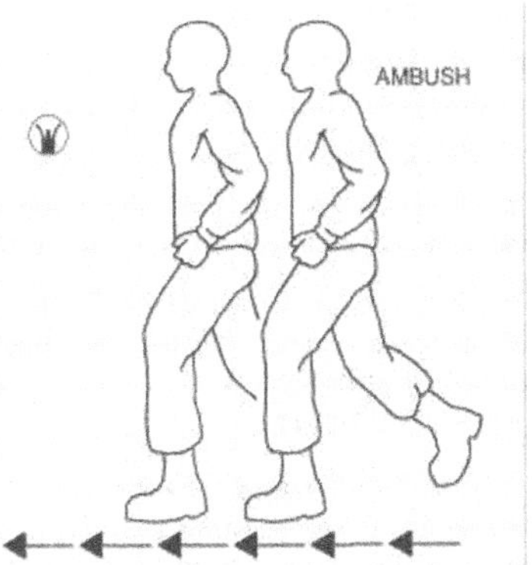

Figure 10-9. Ambush Method.

Trigger Control

As with any target engagement, trigger control is critical to the execution of shots. A Marine can apply pressure on the trigger prior to establishing sight picture, but there should be no rearward movement of the trigger until sight picture is established. Interrupted trigger control is not recommended because the lead will be lost or have to be adjusted to reassume proper sight picture. When using the tracking method, continue tracking as trigger control is applied to ensure the shot does not impact behind the moving target.

Follow-through

If a Marine uses the tracking method to engage moving targets, he continues to track the target during follow-through so the desired lead is maintained as the bullet exits the muzzle. Continuous tracking also enables a second shot to be fired on target if necessary.

Stable Position

To engage moving targets using the tracking method, the rifle must be moved smoothly and steadily as the target moves. A stable position steadies the rifle sights while tracking. Additional rearward pressure may be applied to the pistol grip to help steady the rifle during tracking and trigger control. Elbows may be moved from the support so the target can be tracked smoothly.

10008. Engaging Targets at Unknown Distances

Hasty Sight Setting

While a BZO is considered true for 300 meters, a Marine must be capable of engaging targets beyond this distance. The rifle's sighting system allows sight settings for distances out to 800 meters in 100-meter increments. If a Marine must establish a BZO for extended ranges, it is referred to a hasty sight setting. To achieve a hasty sight setting, a Marine dials the appropriate range numeral on the rear sight elevation knob that corresponds to the range to the target. For example, if the rear sight elevation knob is set at 8/3 and a target appears at 500 meters, rotate the knob to the 5 setting.

Note

Upon completion of firing with a hasty sight setting for extended ranges, return the rear sight to the BZO setting.

Point of Aim Technique

If the distance to the target is beyond the BZO capability of the rifle and time does not permit adjustment of the sights, a Marine can use offset aiming techniques to apply a point of aim for elevation to engage the target. See paragraph 10003.

10009. Engaging Targets During Low Light and Darkness

Combat targets are frequently engaged during periods of darkness or under low-light conditions. Although basic marksmanship fundamentals do not change, the principles of night vision must be applied and target detection is applied differently. During periods of darkness or low light, a Marine's vision is extremely limited. A Marine must apply the techniques of night observation in order to detect potential targets, and he must develop skills that allow him to engage targets under these conditions.

Night Vision

A Marine can improve his ability to see during periods of darkness or low light by obtaining and maintaining night vision. Since adapting to night vision is a slow and gradual process, steps should be taken to protect night vision once it is obtained.

Obtaining Night Vision

There are two methods used to obtain night vision. The first method is to remain in an area of darkness for about 30 minutes. This area can be indoors or outdoors. The major disadvantage of this approach is that an individual is not able to perform any tasks while acquiring night vision in total darkness. The second method is to remain in a darkened area under low intensity red light (similar to the light used in a photographer's darkroom) for about 20 minutes, followed by about 10 minutes in darkness without the red light. This method produces almost complete night vision adaptation while permitting the performance of some tasks during the adjustment period.

Maintaining Night Vision

Because the eyes take a long time to adjust to darkness, it is important to protect night vision once it is acquired. To maintain night vision—

- Avoid looking at any bright light. Bright light will eliminate night vision and require readaptation.
- Shield eyes from parachute flares, from spotlights or from headlights.
- When using a flashlight to read a map or any other written material:
  - Put one hand over the glass to limit the area illuminated and the intensity of the light. Keeping one eye shut will reduce the amount of night vision lost.
  - Cover the light with a red filter to help reduce the loss of night vision.
  - Minimize the time spent using a flashlight.

Factors Affecting Night Vision

Some physical factors may affect your night vision and reduce your ability to see as clearly as possible in low light or darkness. These factors include—

- Fatigue.
- Lack of oxygen.
- Long exposure to sunlight.
- Heavy smoking.
- Drugs.
- Headaches.
- Illness.
- Consumption of alcohol within the past 48 hours.
- Improper diet.

Searching Methods

Once night vision has been acquired, the Marine can locate targets. Some daylight observation techniques (e.g., searching for target indicators) also apply during periods of darkness or low light.

Off-center Vision

Off-center vision is the technique of keeping the attention focused on an object without looking directly at it (see fig. 10-10). To search for targets using off-center vision, never look directly at the object you are observing. You will see the object much better by using off-center vision. Look slightly to the left, right, above or below the object. Experiment and practice to find the best off-center angle for you. For most people, it is about 6 to 10 degrees away from the object, or about a fist's width at arm's length.

Note

Staring at a stationary object in the dark may make it appear to be moving. This occurs because the eye has nothing on which

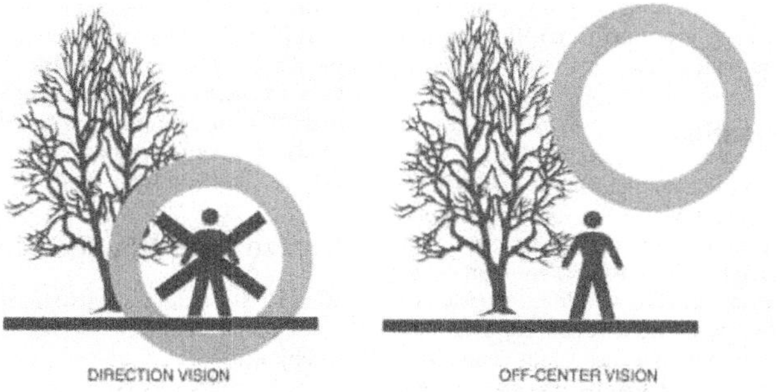

Figure 10-10. Off-center Vision.

to reference the exact position of the object. This illusion can be prevented by visually aligning the object against something else, such as a finger at arm's length.

Scanning/Figure Eight Scan

Scanning is the use of off-center vision to observe an area or object and involves moving the eyes in a series of separate movements across the objective area.

A common method of scanning is to move the eyes in a figure eight pattern (see fig. 10-11). A Marine moves the eyes in short, abrupt, irregular movements over and around the area. Once a target indicator is detected, focus is concentrated in that area, but not directly at it. Pause a few seconds at each point of observation

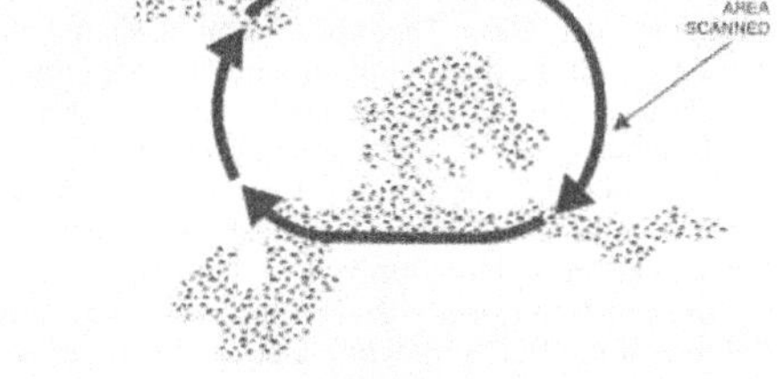

Figure 10-11. Figure Eight Scan.

since the eyes cannot focus on a still object while in motion. Rest the eyes frequently when scanning.

While you are observing, there may be periodic blackouts of night vision due to simple fatigue. This is normal and is not a cause for alarm. Night vision will quickly return after the eyes are moved and blinked a few times.

It is more effective to scan from a prone position or a position closer to the ground than the object being observed. This creates a silhouetted view of the object.

When scanning an area, look and listen for the same target indicators as in daylight: movement, sound, and improper camouflage.

Objects in bright moonlight/starlight cast shadows just as in sunlight.

Sound seems louder at night than during daylight.

Types of Illumination

There are two types of illumination that assist engagement during low light or darkness: ambient light and artificial illumination. Both ambient light and artificial illumination can affect target perception (distance and size) and night vision capabilities.

- Ambient light is the light produced by natural means (i.e., the sun, moon, and stars). Variations occur in ambient light due to the time of day, time of year, weather conditions, terrain, and vegetation.
- Artificial illumination is the light produced by a process other than natural means. Artificial light can be used to illuminate an area for target detection

or to illuminate a specific target to pinpoint its position. There are two types of artificial illumination used in combat: air and ground.

Effects of Illumination

In some combat situations, ambient light and artificial illumination may assist a Marine in locating targets. However, this light can affect perception of the target and disrupt night vision. The introduction of artificial light requires the eyes to make a sudden, drastic adjustment to the amount of light received. This can cause a temporary blinding because night vision was abruptly interrupted. Ambient light also can cause a blinding effect; e.g., a Marine may experience temporary blindness or reduced night vision if a bright moon suddenly appears from behind the clouds.

- Light behind a Marine or light between the Marine and a target illuminates the front of the target and makes it appear closer than it is.
- Light beyond the target displays the target in silhouette and makes it appear farther away than it is. If the target is silhouetted, it is easier to see and easier to engage.
- Air illumination devices are in constant motion as they descend to the ground. This movement creates changing shadows on any illuminated target causing a stationary target to appear as if it is moving.

10010. Engaging Targets while Wearing the Field Protective Mask

While engaging targets in a combat environment, a Marine is under considerable stress caused by fear, fatigue, and the noise of battle. His stress is further aggravated by the fear and uncertainty associated with a nuclear, biological, and chemical (NBC) threat. However, a Marine must be able to operate under any battlefield condition, including an NBC environment. If a Marine wears the field protective mask, its bulk and reduced visibility can affect his firing position which in turn affects the rifle's zero and his ability to engage the target. A Marine must make adjustments to his firing position and the application of marksmanship fundamentals to counter the additional gear worn in an NBC environment. Therefore, a Marine should practice wearing his field protective mask when he is not in a combat environment. This allows him to overcome any restrictions caused by the mask, develop confidence in his ability to execute well-aimed shots while wearing the mask, and develop a plan of action. This plan should address how the rifle is presented to the target, how long the mask is worn, and the likelihood of enemy contact. If a Marine expects to wear the mask for an extended period and enemy contact is likely, he should consider adjusting the rifle sights so that his first rounds are on target.

Marksmanship Fundamentals

Wearing the field protective mask requires a Marine to make modifications to his aiming and breath control techniques.

Aiming

Wearing the field protective mask affects the aiming process and the ability to locate targets. The bulk of the mask may require an adjustment to stock weld, eye relief, head position, and placement of the buttstock in the shoulder.

Breath Control

Wearing the field protective mask affects breath control because breathing may be more difficult. Temporary fogging of the lens also may be experienced. If fogging occurs, a Marine should take a deep breath and fire while holding a full breath of air (inhaling clears the fog).

Firing Position

A good firing position provides balance, control, and stability during firing. The field protective mask's added bulk and other restrictions may require a Marine to make changes to his firing position. The adjustments are unique to each Marine and based on his body size and shape and his ability to adapt to the mask. Adjustments should be minor. However, all firing positions will be affected in the following areas:

Stock Weld

Changing the placement of the cheek on the stock may affect the rifle's BZO. Therefore, a Marine should obtain a BZO for the rifle in full mission-oriented protective posture gear. Stock weld will not be as comfortable or feel as solid as it does without the field protective mask. The loss of sensitivity between the cheek and the stock, due to the mask, may cause the cheek to be pressed too firmly against the stock. Pressing the cheek too firmly against the stock can cause the seal of the field protective mask to break. If this occurs, quickly clear the mask and resume a firing position. If the lens of the field protective mask fogs up while in a firing position, this indicates that the mask's seal has been broken. Clear the mask and resume the firing position.

Eye Relief

The added bulk of the field protective mask may increase eye relief because the head is farther back along the stock. If the eye is too far from the rear sight aperture it may be difficult to acquire the target and to maintain sight picture; if the eye is too close, the rear sight can hit the mask, possibly breaking a lens or its seal.

Head Position

The mask's shape and bulk can make sight alignment difficult to achieve. The restrictive vision caused by the mask may force a Marine to roll or tilt his head over the stock to achieve sight alignment. The Marine should keep his head as erect as possible while maintaining sight alignment.

Placement of Buttstock in the Shoulder

Placement of the buttstock in the shoulder pocket may have to be altered due to the mask's added bulk. If the rifle is canted, a Marine may place the buttstock of the rifle just outside the pocket of the shoulder to achieve sight alignment. Holding the rifle straight is the preferred method of obtaining sight alignment. However, if sight alignment cannot be achieved in this position, a Marine may alter the hold of the rifle to bring the aiming eye in line with the sights. Canting the rifle drastically affects the rifle's zero. A Marine should cant the rifle only as much as is needed to obtain a good stock weld and proper sight alignment. If the rifle is canted, the point of impact may not coincide with the point of aim. For example, when wearing the mask, a right-handed Marine's point of impact is usually high and to the left of center mass (for a left-handed Marine, high and to the right of center mass). Therefore a Marine has to offset aim an equal and opposite distance low and to the right. See paragraph 10003 for a discussion on offset aiming and the known strike of the round technique.

Appendix A

DATA BOOK

Note

The principles for recording data in the data book are the same for the Entry Level Rifle (ELR) Program and the Sustainment Level Rifle (SLR) Program. This appendix provides generic information for completing the data book. Specifics pertaining to each program are contained in the ELR and SLR lesson plans.

1. Data Book

Of all the tools that assist the Marine in firing accurately and consistently, the data book, if properly used, is the most valuable asset. It contains a complete record of every shot fired and the weather conditions and their effects on shooting. When used properly it will assist the Marine in establishing and maintaining a battlesight zero (BZO).

2. Recording Data Before Firing

Recording information in the data book prior to firing saves valuable time on the firing line that should be used to prepare for firing. Some information can be recorded before going to the firing line. In the BEFORE FIRING section of the data book, record the following (see fig. A-1):

True Zero

Front Elevation

Enter the front sight post setting by recording the number of clicks up (+) or down (-) under FRONT ELEV.

Rear Elevation

Circle the rear sight elevation knob setting under REAR ELEV for the yard line firing from—

- 200 yards: 8/3-2
- 300 yards: 8/3
- 500 yards: 5

Note

At 500 yards, the rear sight elevation knob setting may be plus or minus one or two clicks off of 5.

Wind

Under the WIND column under TRUE ZERO, the R represents clicks right on the rifle from the initial sight setting and the L represents clicks left on the rifle. Enter the rear sight windage knob setting by recording the number of clicks right (clockwise) or left (counterclockwise) under WIND.

Wind

Prior to firing, check the wind. If wind conditions are present, a sight adjustment will have to be made prior to firing to ensure shots impact the center of the target.

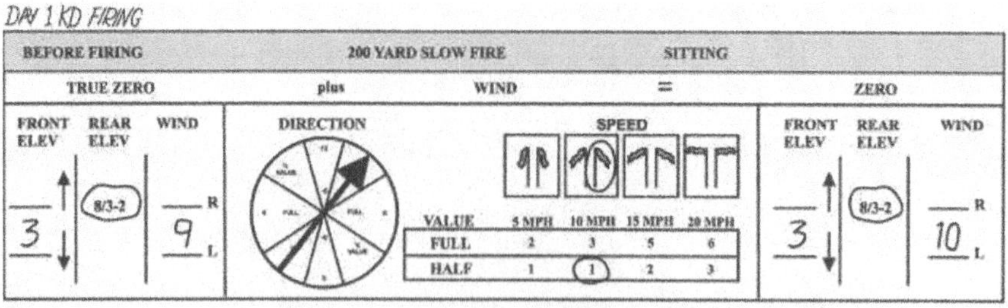

Figure A-1. Recording Data Before Firing.

Direction

Determine the direction of the wind and draw an arrow through the clock indicating the direction the wind is blowing. If the wind is not blowing, a data entry is not needed.

Note

Remember that your position is represented in the center of the clock and the target is at 12 o'clock.

Value

Look at the clock to determine if the wind is full, half, or no value wind. Under VALUE, circle FULL or HALF to indicate the wind value.

Speed

Observe the flag on the range and circle the appropriate flag indicating the wind's velocity (SPEED).

Determining Windage Adjustment

The chart beneath the flag indicates the number of clicks on the rear sight windage knob to offset the effects of the wind. Circle the number of clicks where the wind value and wind speed intersect.

Zero

Determine the zero you will place on your rifle to accommodate wind conditions to begin firing. This ZERO will be the TRUE ZERO plus the rear sight windage setting to compensate for the effects of wind.

Front Elev and Rear Elevation

Elevation adjustments are not affected by wind so the same settings are carried over from TRUE ZERO.

Wind

Wind will affect the strike of the round right or left on the target. Therefore, if wind is a factor, the rear sight windage knob must be adjusted to compensate for the effects of wind.

If the wind is blowing, add the number of clicks circled under the flags to the rear sight windage knob setting from TRUE ZERO.

Once the windage setting is determined, record it in the WIND column.

3. Recording Data During Firing

Recording Data During Slow Fire Stages

The method for calling and plotting slow fire shots in the data book is called "the shot behind method." It allows the Marine to spend less time recording data and more time firing on the target. This is because all the calling and plotting is done while the target is in the pits being marked. This information is recorded in the DURING FIRING portion of the data book page. The proper and most efficient method for recording data during KD slow fire stages is as follows (see fig. A-2):

Fire the First Shot

Fire the first shot. Then immediately check the wind flag to see if the speed or direction of the wind changed. If the wind direction changed, indicate the windage adjustment needed to compensate for it in the WIND row under CALL number 1. This windage adjustment will have to be applied to the rifle prior to firing the second shot.

Call the Shot Accurately

As soon as the shot is fired and the target is pulled into the pits, record the exact location where the tip of the front sight was on the target at the exact instant the shot was fired (assuming sight alignment was maintained). Plot this location on the target provided under number 1 in the block marked CALL.

Prepare to Fire the Second Shot

As soon as you have recorded the call for the first shot, prepare to fire the second shot.

Look at Where the First Shot Hit

As the target reappears out of the pits, look where the first shot hit the target. Remember this location so it can be plotted after firing the second shot.

Fire the Second Shot

Fire the second shot. Then check the wind flag to see if the wind changed speed or direction.

Call the Second Shot and Plot the First Shot

As soon as the second shot is fired and the target is pulled into the pits, record the call of the second shot.

Now plot the precise location of the first shot by writing the numeral 1 on the large target diagram provided in the block marked PLOT.

Prepare to Fire the Third Shot

Repeat steps until all shots have been fired. Indicate each slow fire shot with the appropriate number (e.g., 1, 2, 3, 4, 5).

Make a Sight Adjustment if Required

Sight adjustments should be made off of a shot group, not a single shot. Determine if a sight adjustment is necessary off of the first three shots fired. If the shots form a group (i.e, a group that fits inside the "A" target bull's-eye or the center scoring ring of the "D" target), but are not where they were called, make the necessary sight adjustment.

CAUTION

Generally, major sight adjustments from established sight settings are caused by poor application of the fundamentals, inconsistencies in firing positions, inconsistencies in sight picture at different ranges and different positions, and inconsistent tension on the sling. Every effort should be made to correct shooting errors prior to making a sight adjustment on the rifle.

Note

The plotting targets in the data book for the KD Course of Fire are in inches, requiring you to calculate the number of clicks to center your shot groups.

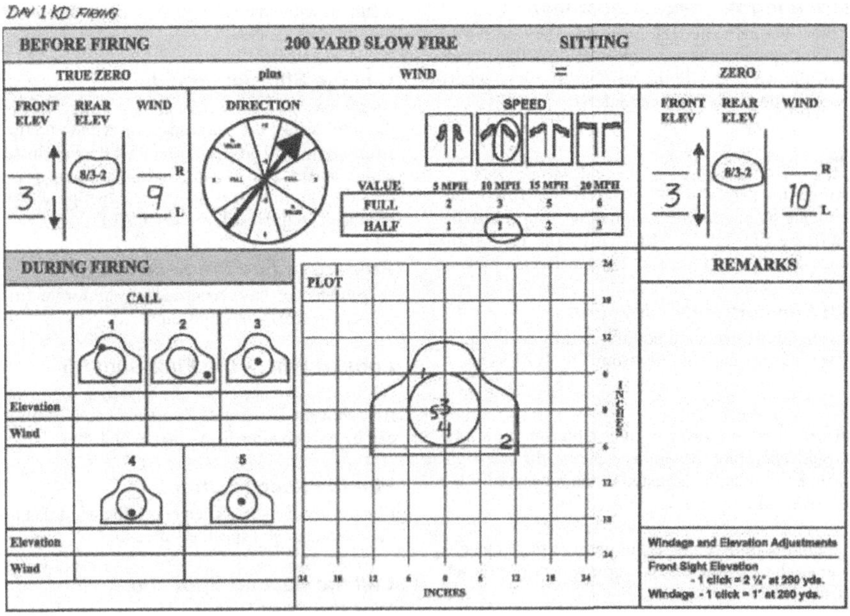

Figure A-2. Recording Slow Fire Data During Firing.

Elevation. Locate the closest horizontal grid line to the center of the plotted shot group. Follow the line across to the numbered vertical scale to determine the number of inches of elevation the shot group is off of target center. Calculate the number of clicks on your front sight post (for 200 and 300 yards) or rear sight elevation knob (for 500 yards) to bring your shot group center.

Windage. Locate the closest vertical grid line to the center of the plotted shot group. Follow the line down to the numbered horizontal scale to determine the number of inches of windage the shot group is off of target center. Calculate the number of clicks on your rear sight windage knob to bring shot group center.

Recording Data During Rapid Fire Stages

In the DURING FIRING section of the data book, record the following (see fig. A-3):

Mentally Call Shots While Firing

While firing the rapid fire string, make a mental note of any shots called out of the group.

Plot the Shot Group

After firing the rapid fire string, and when the target is marked, plot all visible hits with a dot precisely where they appear on the large target diagram in the block marked PLOT.

Make a Sight Adjustment if Required

Locate the center of the shot group. If the shots form a group, make the necessary sight adjustments off of the center of the group. If shots do not form a group and do not just contain a poor shot, do not make a sight adjustment. Determine the sight adjustment by locating the center of the shot group and using the grid lines on the target in the data book. These grid lines represent the number of inches to bring a shot group center. Looking at the shot group:

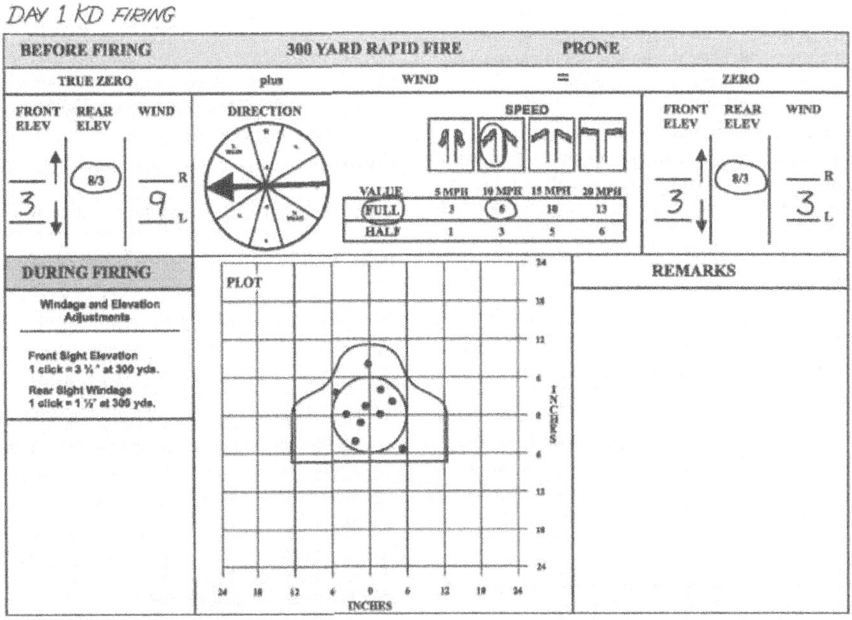

Figure A-3. Recording Rapid Fire Data During Firing.

CAUTION

Generally, major sight adjustments from established sight settings are caused by poor application of the fundamentals, inconsistencies in firing positions, inconsistencies in sight picture at different ranges and different positions, and inconsistent tension on the sling. Every effort should be made to correct shooting errors prior to making a sight adjustment on the rifle.

Note

The plotting targets in the data book for the KD Course of Fire are in inches, requiring you to calculate the number of clicks to center your shot groups.

Elevation. Locate the closest horizontal grid line to the center of the plotted shot group. Follow the line across to the numbered vertical scale to determine the number of inches of elevation the shot group is off of target center. Calculate the number of clicks on your front sight post to bring your shot group center.

Windage. Locate the closest vertical grid line to the center of the plotted shot group. Follow the line down to the numbered horizontal scale to determine the number of inches of windage the shot group is off of target center. Calculate the number of clicks on your rear sight windage knob to bring your shot group center.

Remarks

After firing a stage, record any data or information that can be helpful in improving shooting in the future. Anything done or observed should be recorded. These items will be helpful when analyzing daily shooting performance. Record this information in the REMARKS column. What the Marine fails to record may be the information he will need to improve.

4. Recording Data After Firing

In the AFTER FIRING section of the data book, record the following (see fig. A-4 on page A-6):

Zero

Upon completion of firing, determine the elevation and windage required to center the shot group (if necessary) and record this sight setting in the ZERO block of the AFTER FIRING section. If no adjustments are needed to center the shot group, record the sight settings currently on the rifle.

Front Elevation

Under the column FRONT ELEV, record the final elevation setting made on the front sight post.

Rear Elevation

Under the column REAR ELEV, record the rear sight elevation knob setting for the yard line firing from.

Note

If firing from 500 yards, record the final elevation setting made on the rear sight elevation knob.

Wind

Under the column WIND under ZERO, record final windage setting made on the rear sight windage knob.

Wind

Calculate the prevailing wind during the string of fire.

Direction

Determine the direction of the wind and draw an arrow through the clock indicating the direction the wind is blowing.

Note

Remember that your position is represented in the center of the clock and the target is at 12 o'clock.

Value

Look at the clock to determine if the wind is full, half, or no value wind. Under VALUE, circle FULL or HALF to indicate the wind value.

Speed

Observe the flag on the range and circle the appropriate flag indicating the wind's velocity (SPEED).

Determine Windage Adjustment

The chart beneath the flag indicates the number of clicks on the rear sight windage knob to offset the effects of the wind. Circle the number of clicks where the wind value and wind speed intersect.

True Zero

A true zero is the established zero without the windage adjustments to compensate for the effects of the string of fire's wind. A true zero is calculated because, the next time you fire, the wind conditions will probably be different. Therefore, the rear sight windage knob adjustments made to compensate for the string of fire's wind will not be the correct setting for wind conditions on other strings of fire or other days of firing.

Front Elevation and Rear Elevation

Because elevation adjustments are not affected by wind, the same settings are carried over from ZERO.

Wind

Calculate the windage adjustment to compensate for the string of fire's wind conditions the same way it was calculated in the BEFORE FIRING information of the data book. The only exception is now windage adjustments are being removed from the rifle rather than added to the rifle. Because the windage setting is being removed from the rifle, remove the number of clicks of windage right or left from the ZERO windage setting. Once the windage setting is determined, record it in the WIND column.

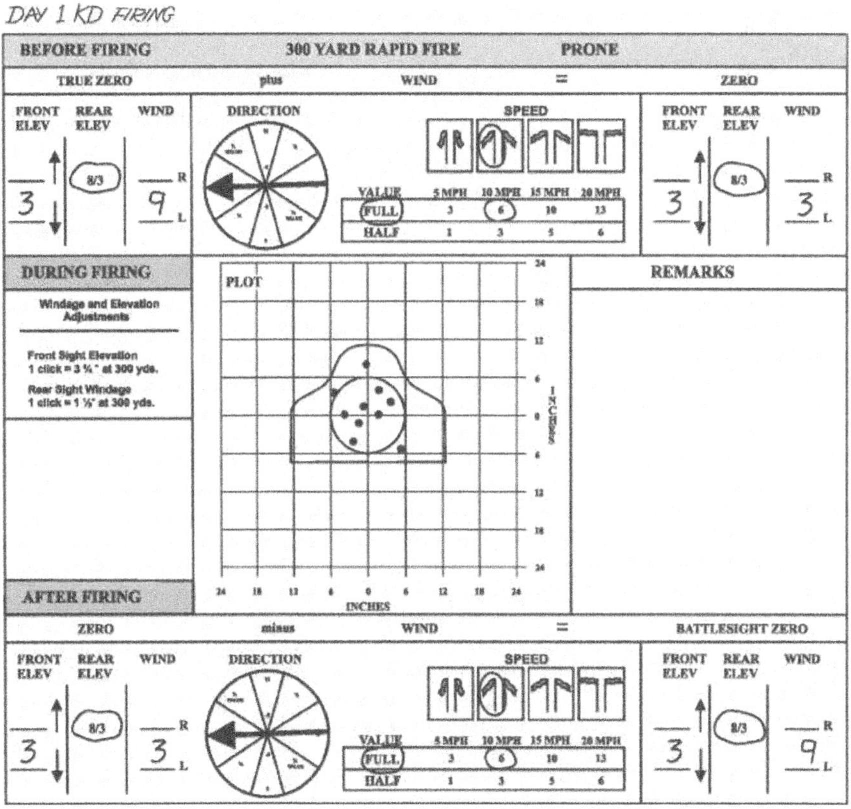

Figure A-4. Recording Data After Firing.

Appendix B

Glossary

Section I. Acronyms

BZO . . . . battlesight zero

CLP . . . . cleaner, lubricant, and preservative

KD . . . . known distance

LAW . . . . lubricating oil, arctic weapons

NBC . . . . nuclear, biological, and chemical

ROE . . . . rules of engagement

SOP . . . . standing operating procedures

TM . . . . technical manual

SECTION II. DEFINITIONS

A

aiming point – The aiming point is the precise point where the tip of the front sight post is placed in relationship to the target.

alibi – Any condition caused by the weapon, ammunition, or range operation that causes the shooter not to have an equal opportunity to complete a string of fire as all other shooters on the range.

B

battlesight zero (BZO) – The elevation and windage settings required to engage point targets from 0-300 yards/meters under ideal weather conditions (i.e., no wind).

bone support – The body's skeletal structure supporting the rifle's weight.

breath control – Procedure used to fire the rifle at the moment of least movement in the body and the rifle.

C

canting – An angular deviation of the weapon to the left or right from a vertical position during firing.

center of mass – A point that is horizontally and vertically centered on the target.

center of mass hold – The placement of the tip of the front sight on the target center of mass prior to the shot breaking.

centerline of the bore – An imaginary straight line beginning at the chamber end of the barrel and proceeding out of the muzzle.

chamber check – Procedure used to determine a weapon's condition.

D

double feed – Attempted simultaneous feeding of two or more rounds from the magazine.

dry fire – Cocking, aiming, and squeezing the trigger of an unloaded rifle in order to practice the fundamentals of marksmanship.

detailed search – Method for conducting a systematic search of an area for specific target indicators.

E

eye relief – The distance between from the rear sight aperture to the aiming eye.

F

function check – Procedure used to ensure the selector lever operates properly.

flag method – Procedure used to determine wind velocity and direction on a Known Distance (KD) range.

field expedient battlesight zeroing – Process used to zero the rifle at 36 yards or 30 meters when a 300-yard/-meter range is not available.

G

gas operated – A self-loading firearm that utilizes the expanding force of the propellant's gases to operate the action of the weapon.

H

hasty search – Method for quickly searching an area for enemy activity.

hasty sight setting – A zero established for distances out to 800 yards/meters.

I

immediate threat target – A target engaged at a short range (50 meters or less) with little or no warning.

initial sight setting – Sight setting placed on a rifle that serves as the starting point from which all sight adjustments are made for the initial zeroing process.

L

limited technical inspection (LTI) – An inspection performed by an armorer on a weapon to determine its operational status (safety and function, not accuracy).

line of sight – An imaginary line extending from the shooter's eye through the rifle's sights and onto an aiming point on a target.

load – Command/Procedure used to take a weapon from Condition 4 to Condition 3 by inserting a magazine with rounds.

M

magazine – A container that holds ammunition in a position to be chambered.

magazine fed – A mechanical, automatic means of supplying a firearm with ammunition to be chambered.

make ready – Command/Procedure used to take a weapon from Condition 3 to Condition 1 by chambering a round.

muscular relaxation – The state of tension required to properly control the rifle. The shooter's muscles are in a relaxed state of control—tightened but not tensed.

N

natural point of aim – The location at which the rifle's sights settle if bone support and muscular relaxation are achieved.

O

observation method – Procedure used to determine wind velocity and direction in a tactical situation.

P

pie technique – A technique used to locate and engage targets from behind cover, while minimizing the Marines exposure to enemy fire.

R

range – **1.** A designated place where live fire weapons training is conducted. **2.** The horizontal distance to which a projectile can be propelled to a specified target. **3.** The horizontal distance between a weapon and target.

recoil management – The ability to manage or control the recoil of the rifle for a given shooting position.

rollout – A technique used to locate and engage targets from behind cover, while minimizing the Marines exposure to enemy fire.

S

semiautomatic – One full cycle of operation. The firing, extraction, and ejection of the spent cartridge, the cocking of the weapon and chambering of the succeeding round of ammunition after the trigger is pulled.

shooter error — Any action induced by the shooter that causes the weapon to fail to operate properly or miss the intended target.

sight alignment – The placement of the tip of the front sight post in the center of the rear sight aperture.

sight picture — The placement of the tip of the front sight post in the center of the target while maintaining sight alignment.

sling – When properly attached to the rifle, the sling provides maximum stability for the weapon and helps reduce the effects of the rifle's recoil. Also used for individual weapon transport.

stability of hold – The ability to acquire a stable position and to hold the rifle steady for any given rifle position.

stock weld – The firm, consistent contact of the cheek with the weapon's buttstock.

stoppage – Any condition that causes the rifle to fail to fire.

T

target indicators – Anything that reveals the enemy's position.

trajectory – The path of a projectile through the air and to a target.

triangulation process – Process used to determine the vertical and horizontal sight adjustments that must be made to center a shot group.

trigger control – The skillful manipulation of the trigger that causes the rifle to fire without disturbing sight alignment or sight picture.

U

unload – Command/Procedure used to take a weapon from any condition to Condition 4.

unload and show clear – Command/Procedure used to take a weapon from any condition to Condition 4 while requiring a second individual to check the weapon to verify that no ammunition is present.

user serviceability inspection – Procedure used to ensure a weapon is in an acceptable operating condition, conducted by the shooter.

V

velocity – Speed at which the projectile travels.

W

weapons carry – Procedure used to effectively handle the rifle while remaining alert to possible threat levels.

weapons condition – Describes a weapon's readiness for live fire.

weapons transport – Procedure used to carry the rifle for long periods of time and when one or both hands are needed for other work.

windage and elevation rules – Rules that define how far the strike of the round will move on the target for each click of front and rear sight elevation or rear sight windage for each 100 yards/meters of range to the target.

Z

zeroing – The process used to adjust the rifle sights that cause it to shoot to point of aim at a desired range.

Made in United States
North Haven, CT
04 November 2026

21774403R00143